# Data, Voice, and Video Cabling

# DATA, VOICE, AND VIDEO CABLING

JAMES HAYES

AND

PAUL ROSENBERG

DELMAR

THOMSON LEARNING

Africa • Australia • Canada • Denmark • Japan • Mexico • New Zealand • Philippines
Puerto Rico • Singapore • Spain • United Kingdom • United States

## NOTICE TO THE READER

Publisher does not warrant or guarantee any of the products described herein or perform any independent analysis in connection with any of the product information contained herein. Publisher does not assume, and expressly disclaims, any obligation to obtain and include information other than that provided to it by the manufacturer.

The reader is expressly warned to consider and adopt all safety precautions that might be indicated by the activities herein and to avoid all potential hazards. By following the instructions contained herein, the reader willingly assumes all risks in connection with such instructions.

The Publisher makes no representation or warranties of any kind, including but not limited to, the warranties of fitness for particular purpose or merchantability, nor are any such representations implied with respect to the material set forth herein, and the publisher takes no responsibility with respect to such material. The publisher shall not be liable for any special, consequential, or exemplary damages resulting, in whole or part, from the readers' use of, or reliance upon, this material.

**Delmar Staff**
Publisher: Alar Elken
Acquisitions Editor: Mark Huth
Project Editor: Barbara L. Diaz
Executive Production Manager: Mary Ellen Black
Art and Design Coordinator: Rachel Baker
Editorial Assistant: Dawn Daugherty



Printed in Canada
6 7 8 9 10 ITC 05 04 03 02

For more information, contact Delmar, 3 Columbia Circle, PO Box 15015, Albany, NY 12212-0515; or find us on the World Wide Web at http://www.delmar.com

**International Division List**

**Asia**
Thomson Learning
60 Albert Street, #15-01
Albert Complex
Singapore 189969
Tel: 65 336 6411
Fax: 65 336 7411

**Japan:**
Thomson Learning
Palaceside Building 5F
1-1-1 Hitotsubashi, Chiyoda-ku
Tokyo 100 0003 Japan
Tel: 813 5218 6544
Fax: 813 5218 6551

**Australia/New Zealand:**
Nelson/Thomson Learning
102 Dodds Street
South Melbourne, Victoria
3205
Australia
Tel: 61 39 685 4111
Fax: 61 39 685 4199

**UK/Europe/Middle East**
Thomson Learning
Berkshire House
168-173 High Holborn
London
WC1V 7AA United Kingdom
Tel: 44 171 497 1422
Fax: 44 171 497 1426

**Latin America:**
Thomson Learning
Seneca, 53
Colonia Polanco
11560 Mexico D.F. Mexico
Tel: 525-281-2906
Fax: 525-281-2656

**Canada:**
Nelson/Thomson Learning
1120 Birchmount Road
Scarborough, Ontario
Canada M1K 5G4
Tel: 416-752-9100
Fax: 416-752-8102

**Spain:**
Thomson Learning
Calle Magallanes, 25
28015-MADRID
ESPANA
Tel: 34 91 446 33 50
Fax: 34 91 445 62 18



**Library of Congress Cataloging-in-Publication Data**
Hayes, Jim, 1946-
Data, voice, and video cabling/Jim Hayes and Paul Rosenberg.
p. cm.
ISBN 0-7668-0964-1 (alk. paper)
1. Telecommunication wiring. 2. Telecommunication cables.
I. Rosenberg, Paul. II. Title.
TK5103.12.H39 1999 99-18044
621.382—dc21 CIP

# CONTENTS

# INTRODUCTION

The rapid growth of communications in the past two decades has been fueled by the development and acceptance of the personal computer and fax machine, the digitization of the phone system, and the end of monopolies in telecommunications. Networks, whether they be LANs (local area networks), MANs (metropolitan area networks), or WANs (wide area networks), have grown geometrically in number and capacity over this time period, especially to support the rise in Internet traffic.

Sometimes we forget the growth of these technologies depends on the development of appropriate infrastructures. Telecommunications depends on reliable, high-capacity communications links that must also be cost-effective. Those links have been provided primarily by fiber optics for the backbone and unshielded twisted pair cable to the desktop.

The continued growth of communications networks also depends on recruiting and training adequate numbers of technicians who can install and maintain those networks. During this time period, we have been involved in training—first the training of installers of fiber optics, and now the training of those involved in wire-based networks through the Fiber U®, Wire U®, and Cable U® training programs.

This book is intended as a textbook for training installers of networks, primarily those installing what is called premises cabling, inside buildings or campuses, carrying voice, data, and video signals. It includes virtually no theory, and only basic descriptions of communications networks in general. Instead it focuses on the practical aspects of designing, installing, testing, and troubleshooting cable plants for these networks.

This book has been developed for the training programs of the Fiber U® (fiber optic) and the Wire U® (copper-wiring) training conferences. These programs offer a combination of classroom seminars and hands-on training, which has been carried over into this book. You will find chapters on the technology, then detailed instructions on installing the typical components of a cable plant.

The timeliness of the material is important in any field that is moving rapidly, and that is certainly the case with communications. We have tried to include the latest material as of our publishing date, but as we go to press, we have to leave incomplete issues like Cat 5e, Cat 6, and Cat 7 UTP cable in the copper domain and the acceptance of a new generation of fiber optic connectors and the possible reemergence of 50/125 fiber for LANs. It is always a good idea to read the trade press and contact the vendors of cabling products for the latest applications and product information. Up to date information is always on the Cable U website: www.CableU.net.

The authors welcome your feedback on how we can improve this book in future editions and make it more useful to you.

Although every precaution has been taken in the preparation of this book, the authors and the publisher assume no responsibility for errors or omissions. Neither is any liability assumed for damages resulting from the use of information contained herein.

## A NOTE OF APPRECIATION

We wish to thank all those who contributed to our research for this book by providing us with product and technical information for inclusion, encouraging us in putting the material together, and reviewing our rough drafts. Thanks first to Fotec, Inc., the fiber optic test equipment company and the originator of Fiber U® and Cable U®, for providing much of the supporting material used in this book. We especially wish to thank Jim Shane, our teacher summer intern who helped make this book more understandable, Ruth Rosenberg, who helped with the history section, along with contributors Doug Elliott, Marilyn Michelson, Clay Paniati of Independent Electrical Contractors, Bob Rose of TAD Technical Institute, Jim Bordyn of Datacom Technologies, Tim Berelson of Prestolite, Jeffrey Brown of ICC, Shannon Burke of Paladin Tools, Tom Debiec of Berk-Tek, Michael Mayfield of Remee, and reviewers Clay Laster of San Antonio College, Robert Baird of Independent Electrical Contractors, John Highhouse of Lincoln Trail College, and A.J. Pearson of the NJATC. And thanks also goes to the directors of The Fiber Optic Association and everyone at Fotec and elsewhere who has helped make our Fiber U® and Cable U® conferences a success.

Jim Hayes, Medford, Massachusetts
Paul Rosenberg, Chicago, Illinois

# PART 1

# INTRODUCTION TO DATA, VOICE, AND VIDEO CABLING

CHAPTER

# 1

# THE HISTORY OF TELECOMMUNICATIONS

The history of telecommunications spans a mere 150 years, from the development of a practical telegraph to today's worldwide fiber optic and satellite networks. Technology often exists long before its practical applications are put into widespread use. Twenty to thirty years often pass before inventions are widely accepted and used, although today it seems only months pass before technical wizardry becomes a household necessity. Today the pace of change seems much more rapid than it did when telecommunications began.

## TELEGRAPHY

Sometime not too long after 1729 when Stephen Gray demonstrated that thread could carry an electrical charge, people started to think that perhaps messages could be sent electrically. Wire was substituted for thread, and the theoretical basis of telegraphy was aided by Hans Christian Oersted's discovery of the principle of electromagnetism (1819) and Alessandro Volta's invention of the electric cell (1820), now called the battery.

By this time, the development of the telegraph was proceeding independently on three continents; but it was not until 1837 that the first patent for communication was issued. The patent went to Cooke and Wheatstone of England for their telegraph of a five-needle system that used a combination of two of the five needles to point to letters of the alphabet. It was simple to operate but required a run of six wires—one for each needle and a return wire—and, because of current leakage and low-output batteries, was limited to a first installation of about a mile. Ensuing developments, those of a simplified two-needle system and a means of increasing current in a line, enabled Cooke and Wheatstone to install a much longer line—thirteen miles, in fact. Meanwhile Gauss and Weber of Germany developed a successful telegraph using the principles of electromagnetic induction (1833).

The American Samuel Morse was known for portrait painting, not inventing, when he began work on his own telegraphic design in 1832. While returning home from Europe by ship, he fell into the company of one Charles Thomas Jackson, a scientist and physicist who had just procured some crude electrical equipment. Their discussions led Morse to begin development of a practical telegraph. Aided technically and financially, he applied for a patent and introduced his telegraph in 1837, and was awarded a patent in 1840. At this time he was also developing his now-famous Morse code, the signals of dots and dashes used to represent letters during telegraphic transmission. In 1843 Morse was granted by Congress $30,000 for an experimental line between Washington and Baltimore. The line opened May 24, 1844, with the well-known exclamation "What hath God wrought!"

## TELEPHONY

Telegraph had given humans the means to transmit impulses that represented letters. When these letters were received and decoded, they provided a way to convey messages over long distances. Naturally, the next step was to consider whether sound might also somehow be electrically transmitted.

Credit for the telephone has been given to Alexander Graham Bell, though many people were working independently on the concept. Bell was actually working on a harmonic telegraph when he became convinced that speech could be transmitted electrically. He knew that to accomplish this, the intensity of electrical current would have to be made to fluctuate the same way that the density of air varies when sound is produced. Bell overcame the problems of the musical telephone by using a fluctuating current rather than a broken one. He applied for a patent for an "electrical speaking telephone" just hours before Elisha Gray filed for a patent for his telephone. Bell was awarded the patent on March 7, 1876, and just three days later his famous words, "Mr. Watson, come here, I want you," were the first to be transmitted and received via telephone.

Many men contributed to telephone improvements, among them Elisha Gray, Thomas Edison, Swedish engineer Ericson, and David Edward Hughes, whose invention of the microphone in 1878 became universally used in telephones.

It is amazing how quickly the use of the telephone spread. Originally the lines were run between two locations that the customer wished to connect. The first switchboard, an experiment, was installed in Boston in 1877. Just four years after the invention of the telephone, there were 54,000 telephones in the United States!

The year 1884 saw the start of commercial service between New York and Boston; and as demand was growing so quickly, the American Telephone and Telegraph company (AT&T), with which we are all familiar, was formed in 1885. Mr. Bell's patents expired in the years 1893 and 1894, resulting in the sprouting of numerous independent phone companies, although AT&T, later to become the parent of Bell Systems, maintained its place as the largest operator of America's phone network.

In the last twenty years, many changes have occurred in telephony, from both a technological and a business viewpoint. Technologies like digital transmission, electronic switching, and fiber optics have produced massive changes in the way the telephone system works. But even greater changes have occurred in the business aspects, including increased competition, the breakup of AT&T, and the subsequent merger of several of the local operating companies. Mergers between telephone and cable television companies have shown that simple telephony is evolving toward a multimedia future.

## WIRELESS: RADIO AND TELEVISION

Following the development of the telegraph and telephone, experiments began to communicate without the wires. Pioneering work from Giuglielmo Marconi on a wireless telegraph led to the development of radio.

An outgrowth of radio, television was a feasible though not yet pragmatic idea at the time of World War I. Two approaches subsequently were taken toward the development of this concept—the use of mechanical scanners and the use of completely electronic systems. Some of the early mechanical scanners were quite successful and were used in early broadcasts. The British telecast the coronation procession of George VI in this manner.

Experimental telecasting was beginning in the United States, and though the technology existed, use was strictly regulated by the government. No commercial stations were allowed, only experimental ones, while waiting for the development of a federal system of standards. When developments progressed far enough to establish such a system and regular service began, two stations were licensed for commercial use. World War II, however, interrupted the emergence of television in both the United States and Britain. After World War II, the Federal Communications Commission (FCC) designated thirteen VHF (very high frequency) channels for commercial use, and when President Truman was inaugurated in 1949, it is estimated that some 10,000,000 people viewed the telecast.

In subsequent years, several adjustments were made to television—improvements in tuning capabilities for both UHF (ultrahigh frequency) and VHF, and sets that could receive both black-and-white and color broadcasts. Satellite TV was introduced in 1960 and became commercially available in 1965, providing an alternative to commercial networks.

Cable television, providing yet another alternative, was originally confined to receiving and retransmitting those programs originating with over-the-air stations. For example, there were some communities that could not receive broadcast stations, either because of distance or shadow areas where the signal was too weak. Community antennas were installed at a remote location, such as the top of a nearby hill, and signals from them were fed to the homes in the area via copper cables. In 1962, however, the privilege to *originate* programming and accept advertising was granted by the FCC, and this and other rulings contributed to the growth and expansion of cable television.

An enormous boon for cable TV was provided via means of communications satellites for domestic television broadcasting. Networking was a natural next step, and HBO was the first of a dozen or so networks that developed by the end of the decade. Sending their signals via satellite to cable operators around the nation, networks flourished and superstations were formed (WTBS in Atlanta, and WGN in Chicago, for example).

Like telephony, cable television has benefited from new technologies. Satellites expanded the networks to worldwide markets, and fiber optics in the form of hybrid fiber-coax (HFC) networks allowed new services such as Internet connections. Now digital television promises high quality pictures and even more channels from which to choose.

The same technology used in cable TV produced low-cost components that could also be used for closed-circuit monitoring for security and other purposes. Even video conferencing, which requires too much bandwidth for conventional phone lines, may be made possible over the Internet through the latest generation of CATV systems.

## COMPUTER NETWORKS

Many developments preceded the computer as we know it today. What we consider the beginning of real computers is the development of computers during World War II to calculate ballistics. The 1950s computers, called the *first-generation computers*, are those vacuum-tube computers that used the idea of stored programming. The UNIVAC I (Universal Automatic Computer I) was the first mass-produced first-generation computer.

Second-generation computers made use of the transistor, which replaced vacuum tubes. The advantages of the transistor were many—the computers using the transistor were smaller, faster, more reliable, and used less electricity and therefore produced less heat. Third-generation computers, such as the IBM (International Business Machines) System/360, were introduced in 1964 and made use of the integrated circuit. Continued improvements upon this single piece of silicon, on which were placed transistors that had been considerably reduced in size, brought about a corresponding reduction of computer size and cost throughout the 1960s and 1970s.

New developments are being made all the time, as anyone who has just purchased a new PC (personal computer) will understand. Still, many of these developments, which the modern PC user often takes for granted, at one time were revolutionary, or nearly so. Improved methods for data storage and for input and output, continued improvements in microelectronic circuitry, advances in computer language, and the development of floppy disks, specialized programs, and the mouse and icons are just a few of these ongoing developments that have permitted computers to play an active role in the daily lives of a large percentage of our population.

Computers can be linked by several methods. Linking allows the sharing and transferring of information, thereby multiplying the resources of the single computer, and can be accomplished through directly linking two computers with a cable, through a *LAN* (local area network), or over the phone, using a modem. Ethernet, the most popular LAN, was developed by Bob Metcalfe and fellow researchers at Xerox Palo Alto Research Center in the early 1970s, starting with a drawing on a napkin!

When linked with a LAN, computers can share common equipment, such as printers, and send e-mail (electronic mail) and files to other persons linked by the same LAN. Two or more LANs may be linked to form an even larger network. In some instances, even databases can be shared. It is necessary to maintain the shared files and control the shared equipment, and this is done through a *network server*, or *file server*, typically one or two personal computers that function in this capacity.

The biggest recent development in communications has been the Internet. The Internet is a worldwide network of computer networks that allows computers anywhere to exchange messages, files, graphics, and so on. With the development of software to support publishing documents on the Internet and browsers to allow accessing it easily, the World Wide Web (WWW or "the Web") has become the most widely talked about and most often used communication method today. With proper equipment, the Web can handle voice, data, and video equally well. Even CATV companies are getting into the Internet, offering high-speed connections through cable modems, which provide Internet access by multiplexing the data over cable as a channel of the CATV system.

With a modem and the proper software, any computer can be linked over the Internet with another computer equipped in the same fashion. Such linkage allows access to telecommunication services by which computer operators can send and receive e-mail, access services for on-line information, "chat" with other subscribers, and so on. Stocks, groceries, and houses are sold this way; libraries are accessed; and notices are posted on public bulletin boards. News and weather are available in an up-to-date manner without having to wait for the 6 o'clock news and without having to listen to news in which one might not be interested. Messages can be sent reliably, nearly instantaneously, inexpensively, and, with the proper encoding, with complete privacy between any computers in the world linked and equipped in the necessary fashion. We have come a long way from communication by smoke signals.

The communications community has become obsessed with "convergence," where traditional telephone companies offer Internet connections and video, CATV companies offer Internet and voice phone services, and even local utilities offer communications links along with electricity and gas. This is leading to the requirement to "connect" every computer on a network and the Internet. For the contractor, this means that practically every office and home eventually will have cabling for voice, data, and video, just like electrical wiring.

## CHAPTER REVIEW

1. What prompted the development of the telegraph?
2. For what was Samuel Morse better known than the invention of the telegraph?
3. Who missed the patent for the telephone by being a few hours later than Alexander Graham Bell?
4. What was the original name for radio that is now used for all forms of radio communications?
5. What electronic components characterized first-, second-, and third-generation computers?
6. Ethernet, the most popular computer network, was invented at a company known for something entirely different. What company was it?

CHAPTER 2

# THE TECHNOLOGY OF COMMUNICATIONS

We will be covering three primary communication technologies in this book—voice, video, and data. First, we will cover basic telephone technologies and telephone networks, which (aside from telegraphy) were the first communication technologies, and which provided a technical base for succeeding technologies. Next, we will cover television and video systems, obviously one of our most important technologies. And finally, we will cover computer networks, the importance of which is hard to overestimate.

As we begin this section of the book, we will cover just the basics of these technologies. We will then progress to cabling technologies, first covering copper wire and then fiber optics, with complete explanations of their operation and installation.

## THE OPERATION OF TELEPHONES

Modern telephones operate on essentially the same principles that were developed over 100 years ago. They use a single pair of wires that connects the phones and a power source. When phones are connected, the power source causes a current to flow in a loop, which is modulated by the voice signal from the transmitter in one handset and excites the receiver in the opposite handset (Figure 2–1).

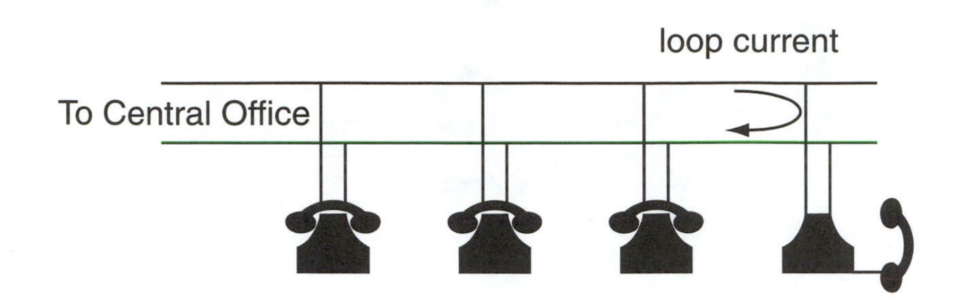


**Figure 2–1.** Telephones are connected in parralel and operate on a current loop.

Dialing was originally done by a rotary dial that simply switched the current on and off in a number of pulses corresponding to the number dialed. Dialing is now mostly (but not entirely) accomplished with tones.

By operating on a current loop, phones can be powered from a central source and extended simply by adding more wire and phones in parallel. Most phones are now electronic (that is, they use semiconductors rather than electromechanical devices), but they use the same type of wiring, frequently called *current-loop wiring*. In office systems, the phones are sometimes digital, and they use twisted pair wiring in a manner similar to computer networks, but at lower speeds.

Telephone wiring is simple because the *bandwidth* of telephone signals is low, generally around 3,000 *hertz* (cycles per second). Bandwidth is similar to speed—a low bandwidth requires lower frequencies, and a higher bandwidth requires higher frequencies. Computer modems use sophisticated modulation techniques to send much faster digital signals over low-bandwidth phone connections.

Because of the low-bandwidth and current-loop transmission, telephone wire is easy to install and test. It can be pulled without fear, and if it is continuous, it should work.

Outside the home or office, the phone connects into a worldwide network of telephones (Figure 2–2), all connected together by a combination of copper wire and fiber optic cables leading to switches that can interconnect any two phones for a voice conversation or data transmission. These switches create a "switched-star" network, where each switch knows how to find every phone by connecting through successive switches. When a phone connection is made, the connection stays complete as long as the users require, creating a continual virtual pathway between the two phones.

Telephone cabling is simple 4-wire cables inside the home or 4-pair unshielded twisted pair (UTP) cables inside an office. The "subscriber-loop" connections to local switches are on multipair wires, with up to 4,200 pairs per cable. Most large pair count cables are being replaced by fiber optics, since those 4,200 pairs of cop-

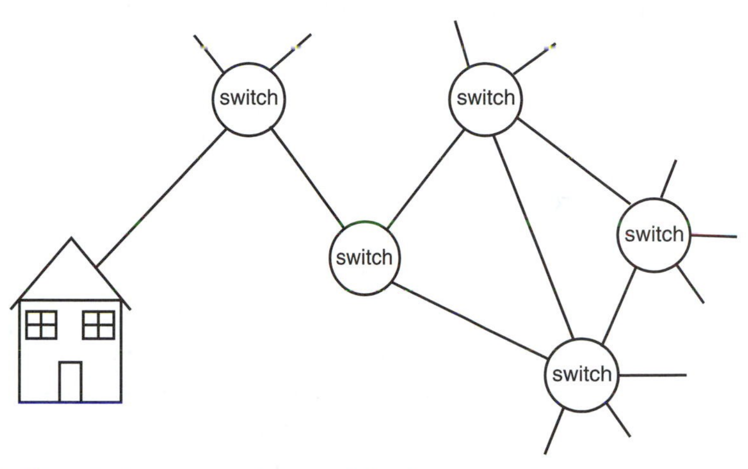


**Figure 2–2.** The telephone network is a switched-star network.

per wires can be replaced easily by a pair of single-mode optical fibers. All fibers used in telephone networks are single-mode types.

Telephone connections between local switches, the "interoffice trunks," are predominately optical fiber already, and soon will be all fiber. Long distance has been virtually all fiber for years, except in areas of exceedingly rugged terrain or isolation, where terrestrial radio or satellite links are more cost-effective. Even fiber optic undersea links have overtaken satellites for international calls because of the higher capacity and lower cost of fiber compared to satellites.

## VIDEO TRANSMISSION

Probably the easiest way to explain basic video or TV technology is to use the fax machine as an example. Fax machines could be called *slow-scan television.* They send one picture frame at a time, in black and white, over telephone lines. We are all familiar with the process: the machine scans one line at a time, interprets that area as black or white space, translates the information into electronic pulses, and transfers the information to a receiver.

Black-and-white television uses the same technology as the fax machine, except at sixty frames per second, and using a camera rather than a scanner. Each line of the image is scanned, and translated into a complex video signal. The color process is the same, except a still more complex signal is used.

Because of the complexity of the television signal, it requires a lot of bandwidth. A standard TV channel uses 6 *MHz* (millions of cycles per second) of bandwidth. In other words, the signal requires six million cycles per second (each cycle contributing part of the signal information) to get all of the information sent fast enough. We could send video images at lower speeds, but the images would be in slow motion.

Closed-circuit TV (CCTV) used for surveillance in security systems consists of only one channel operating with a direct cable connection to a monitor. In large systems, video switches scan numerous cameras to allow the monitoring of many locations, but each remote camera is connected to the central monitor by an individual cable.

In order to use multiple channels, each channel must use separate frequencies, or else their signals would all be jumbled together. For example, the FCC has assigned channel 2 the frequency range of 54 to 60 MHz, channel 3 has 60 to 66 MHz, channel 4 has 66 to 72 MHz, and so on. (UHF channels go up to 890 MHz.) Each channel transmits its programming in that 6-MHz channel.

Broadcast TV sends its signals over the air, by modulating a signal on a transmitting antenna that is received by a smaller antenna on your home or TV. Cable TV works the same way, except it is sending the signal through coaxial copper cable or optical fiber, not via radio waves.

One of the very intimidating words used in the TV business is *modulation.* Modulation is simply the function of taking a basic 6-MHz TV signal and modifying it so that it matches one of the standard channel frequencies.

For example, the TV signal coming out of the camera at your local station operates at about 6 MHz. If we assume that your station is channel 13, it must send its programs in the 210 to 216-MHz slot. So, before sending the signal to the antenna, it must change the 6-MHz signal to a 210- to 216-MHz signal.

This is done with a special electronic device called a *modulator.* Your VCR has a modulator built into it. Since videotape cannot record the very high frequencies used in broadcasting (in the last example, the station used 216 MHz), the VCR must record at frequencies that are too low for a television tuner to pick up. When you play a videotape, the information from the tape is modulated to channel 3 or 4, which your TV can then display.

Once we begin to customize television systems in the home or office, the modulator becomes a very important device. It is necessary to assign the various video inputs to specific channels, so that a television can use them correctly. Video distribution systems allow for custom television channels to be transmitted through the facility. This is generally done in combination with a cable TV system. While these systems are capable of handling well over 100 stations, very few use more than 60 or 70. This allows the owner to transmit other signals through the unused channels. Channels are assigned with the use of a modulator, which applies the basic television signal to any channel.

Sending television signals requires much more bandwidth than that required by telephones. When the technology was first developed, the only cable capable of carrying these high-speed signals with adequately low loss was coaxial cable, also called *coax*. Coax (Figure 2–3) uses a central conductor surrounded by an insulator, then an outer conductive webbing called the shield, and finally a plastic jacket.

It is the design of the cable, with the central conductor widely and evenly separated from the outer conductor, that gives it the high bandwidth capability. In addition, the outer conductor acts to contain the signal inside the cable, reducing the emissions from the cable that cause interference in other electronics and the interference of other outside sources on the signal in the cable itself.

The signal in coax cable is a simple voltage. It can be introduced into the cable at either end, or even in the middle, where it will be carried to either end. In any application, it is important that the transmitters be selected to be appropriate to the characteristics of the coax cable and both ends must be terminated to prevent reflections that can cause interference.

Coax cable must be installed with care. It should not be stretched or kinked. Doing so will reduce the level of bandwidth it will transmit. Connectors should be carefully installed to prevent signal *leakage* (the TV term analogous for signal loss), and unused ends must be properly terminated to prevent reflections.

## CATV NETWORKS

Television is delivered by broadcasting, or simply delivering the same signal to all directions and users at once. It may be delivered by terrestrial antennas to home televisions, sent to satellites that rebroadcast it to large geographic areas at once, or transmitted by coaxial cable on a community antenna television (CATV) system.

Most CATV systems get their broadcast signals from satellites at a location called the headend, which then send it out over large, low-loss coaxial cable, called RG-8, about an inch in diameter. These cables terminate in amplifiers that rebroad-

**Coaxial Cable Construction**

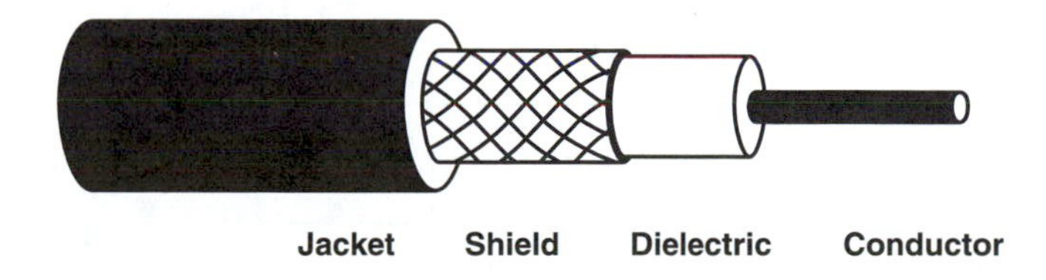


**Figure 2–3.** Video transmissions require the high bandwidth capability of coaxial cable.

cast it over many other cables to more amplifiers, which repeat the rebroadcast. Thus the CATV system (Figure 2–4) looks like a "tree-and-branch" system, the name used to describe this network architecture.

The obvious fault of this type of system is its vulnerability to the failure of a single amplifier. If any amp fails, service is lost to all downstream customers. The solution is to separate the network into smaller segments to prevent massive failures.

Today, most CATV system operators are switching over to a hybrid fiber-coax (HFC) network (Figure 2–5), with fiber optic cable distributing headend signals to local drop amplifiers for distribution to homes. By using fiber for the headend connection, the number of amplifiers connected in series is reduced to about four, minimizing the number of drops affected by any single amplifier or cable fault. This greatly enhances system reliability and minimizes maintenance costs.

Fiber is added to CATV systems already in place by a fiber "overbuild" where the fiber optic cable is lashed to aerial coax or pulled alongside it in ducts. Large systems use fiber to distribute signals to several towns, since fiber has the capability of transmitting long distances, saving on the costs of additional headends.

Once the CATV signal is split off to the house, it is carried on smaller coax cable to and throughout the home to connect all the TVs. This cable, although smaller, can still carry GHz bandwidth signals necessary for transmission of up to 100 or more channels of video programming.

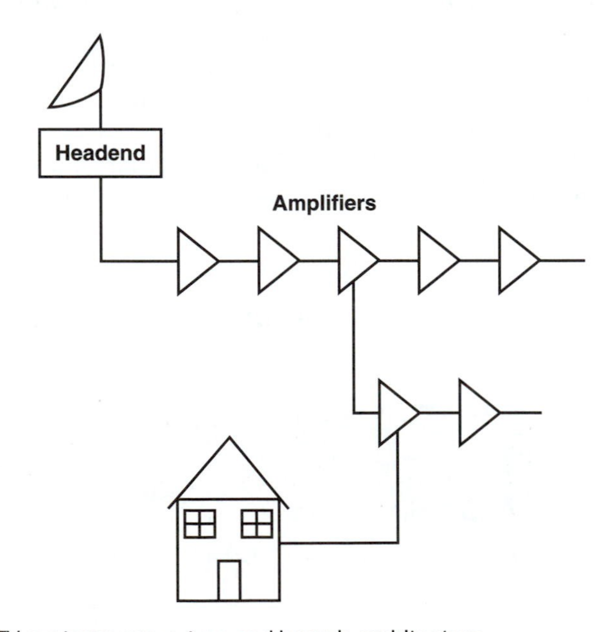


**Figure 2–4.** CATV systems are a tree-and-branch architecture.

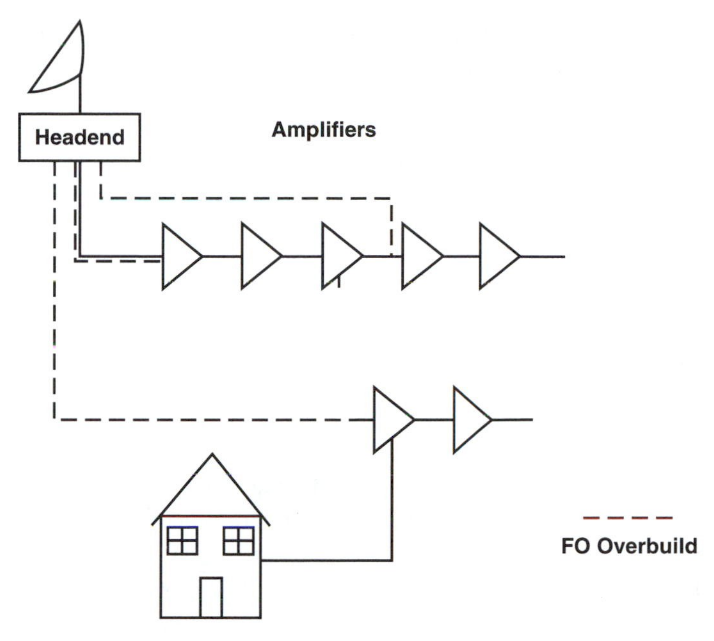


**Figure 2–5.** CATV systems use hybrid fiber-coax (HFC) networks to minimize the number of amplifiers between the headend and subscriber.

## COMPUTER NETWORKS

There is no computing trend in recent years that has been as large or important as networking. The use of networking has risen steadily in the past ten years and is continuing to spread through the marketplace. There are so many different types of networks available, so many variations of connections, variations of signals, and so on, that the entire subject can be very confusing. Our coverage of networking will proceed slowly and carefully, covering and explaining all of the various parts.

The purpose of networking is to allow a number of computers to operate together, sharing information and allowing the operator of one computer to read and use programs in another computer.

Networking is an important technology in offices. For instance, bookkeepers do all of their work on computers in their offices, but unless their computers are somehow connected to their bosses' computers (through a network), the bosses have no way of obtaining financial information on their own company unless the bookkeepers bring them reports. Neither can anyone else in the company get pertinent financial information from the bookkeeping system without a special effort. The same goes for any type of order processing system, sales reports, analysis reports, or whatever a business may use. With a network, however, all of this information (except information that is specifically excluded, such as confidential financial data) is available to whoever needs it, whenever they need it.

The confusing part of networking comes when we devise methods for connecting these computers together and give new names to the various parts that are used to connect them.

The following are the fundamental parts of networks.

### Digital Computers

Phones and TV are basically analog transmission networks. They begin as electrical signals of continuously varying voltages or currents traveling over the cables (Figure 2–6). Computers, however are digital. Their data is represented by "ones" and "zeroes" that can be stored and manipulated by digital processing. Computer programs express all data, graphics, and programs as digital data.

Analog signals are subject to distortion and degradation by noise and attenuation. Digital signals are immune to noise and tolerate greater attenuation, simply because the receiver only needs to distinguish between a one or zero, not a continuously varying signal. Besides computers, most phone signals are digitized as soon as they reach the phone company central office. This not only allows for better signal quality over long distances, but allows for the multiplexing or compressing of digi-

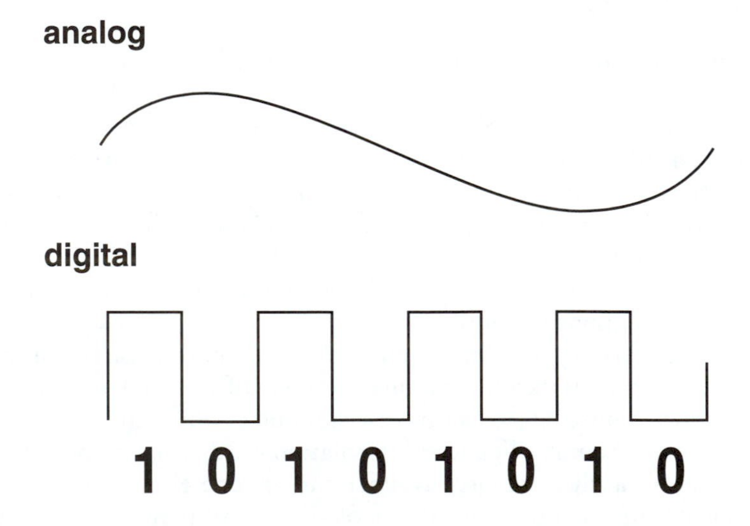


**Figure 2–6.** Analog systems transmit data as a continually varying signal, while digital systems transmit data as a stream of ones and zeros.

tal data into high-bandwidth signals carrying numerous phone calls. Digital television is also in the near future, for similar reasons.

### Computer Connection to the Network

If the network is going to allow personal computers to communicate, each computer must have some way of getting into it. This is done through the use of a *network interface card* (NIC). The NIC is a circuit card that fits into one of the expansion slots in the back of a personal computer. This is the most vital link between the processor and the network. It interprets information between the computer and the network, and feeds information in and out of the computer in a way that both the computer and the network can accept.

### Network Architecture and Protocol

How are the computers linked and what language do they use to communicate? Is one of the processors (computers) the basis of the network (and hence, the most important), or are all processors connected to the network considered and treated equally? Most networks use one computer (usually the most powerful) as the heart of the network, and the other computers on the network use it as their main source of data storage and retrieval. The central computer is generally called the *server,* or *file server,* and the satellite computers are commonly called *clients* or *nodes*. If all computers on a network are equal and share data among themselves, the network is called a *peer-to-peer* network.

### Servers and Hosts

Most networks are *client-server* networks, where a central computer (server) acts as the main depository of all the files that may be shared by all users (clients).

When we are discussing file servers being the center of most networks, it is important to talk about what types of computers these servers are. Typically, when the word *server* is used, it refers to a high-quality personal computer or workstation, usually a new, state-of-the-art unit. When, however, you hear the word *host,* this refers to a *mainframe* or *minicomputer* being used as the center of the network.

Host computers are usually used in large networks at large companies. These companies frequently have large centralized operations that require huge amounts of storage and processing capabilities that are available only in mainframe computers. For these companies, it simply makes sense to connect their networks to this central information source that is already running most of their company's operations.

Small companies, with their different needs, rarely require a mainframe computer. They nearly always use file servers, rather than host computers, or peer-to-peer networks.

### Networking Software

Just like a computer has an operating system that controls all its hardware and allows applications software to run on the computer, networks have operating systems that control the flow of information on the networks. The network operating system establishes what computers are on the network, what peripherals they share, and how data is formatted for transfer, and it even detects and corrects errors in data caused during transmission.

Since networks send various routing commands through their system, software that is not written for networks (where it will be exposed to these strange commands) often will not work properly. Many of the more popular types of software come in a "network version." It is very important to verify that your software is compatible to your network before you install it. Copyright laws may also limit the use of software on networks, depending on the license agreement of the manufacturer.

### Network Protocols

This is the language that computers in a network use to communicate. It specifies how the data is encoded for transmission, how addresses are specified, and how access to the network is gained. Each network is different in these aspects, requiring *bridges* to translate protocols to allow different networks to communicate.

### Network Topology

This is the connection pattern of the computers in the network. There are several methods for connecting all of the computers on the network together. Some of the most common methods are:

1. Star—all computers attached to a central computer or hub
2. Ring—all computers connected in series along a ring
3. Bus—all computers connected to a single cable

A bus topology looks like all of the computers are connected to one central cable. This is how it was actually done with the first Ethernet networks. The cable was a thick, stiff coaxial cable, and the computer was connected directly to the cable by cutting a hole in the cable with a "coring tool" and attaching a network transceiver. Ethernet cable was expensive and difficult to install, but necessary to provide the performance needed for reliable transmission. Later networks used thin coaxial cable, with twist-lock connectors, which was less expensive and easier to install.

In ring topology, the computers are simply connected together in a large ring. This is done with either fiber optic, twisted pair, or coaxial cables, but of the two most popular ring networks, Token Ring uses a shielded 2-pair cable and FDDI (Fiber Distributed Data Interface) uses fiber optics. Note that in this arrangement, each computer has two cables connecting to it, one from the previous computer in line, and one to the next computer in line.

A star topology is an arrangement where all computers on the network connect to a central point. This connecting point can be the file server, or main computer, or it may be a hub or switch. Hubs act as communications points, taking signals from all the connected computers and broadcasting them to all other computers connected to the hub, and then multiplexing the signals to communicate to other computers or hubs over the *backbone*. Switches are similar to hubs, but they do not broadcast inputs; instead, they switch them to their destinations based on the addresses provided by the sending computer.

From this central computer or hub, there are a number of cable runs to the various satellite stations (nodes). In actual practice each computer has only one cable connected to it, with all of these cables feeding into one central location, where the file server or hub will be located.

Which network topology you use is determined by the type of network you plan to install. Some require one type of connection, and some require others. There can be advantages and disadvantages to all types. Networks have evolved, however, to a common star wiring scheme or cabling architecture, no matter what the network topology, to accommodate all topologies.

### Network Cabling

Once you have a network interface in place, you need some method of getting the data signals from one computer on the network to another. This can be done in a number of ways. Although any of the following methods can be used, the first item on the list (twisted pair cables) is far and away the most common:

1. Unshielded twisted pair (UTP) cables.
2. Shielded twisted pair cables
3. Coaxial cables
4. Optical fiber cables
5. Radio waves
6. Infrared light
7. Electronic signals sent through power lines.

The choice of network cabling (or *communications medium*, as it is sometimes called) is rather important because of the extremely high frequencies of the signals. Sending 60-cycle utility power through a wire rarely presents a difficulty; but sending a 100 million bits per second signal can be a little more tricky. For this reason, the method of sending signals, and the materials they are sent through, can be important.

The types of signals that are sent through the network and the speed at which they are sent are extremely important details of a network. All parts of the system must be coordinated together to send, carry, and receive the same types of signals. Usually, these details are not something that you have to consider as long as all parts of your network come from (or at least are specified by) the same vendor or were designed to the same industry standard.

Because networking evolved over several decades, many different cabling solutions have been used. Today, virtually all cabling has moved to UTP or fiber optic cable specified in the EIA/TIA 568 standard. Since older equipment is sometimes used on newer cable plants, adapters called "baluns" have become available to make that adaptation.

## NETWORK EXAMPLES

Ethernet, the most popular network type, started on a large coaxial cable, using a bus structure, where each terminal was connected directly to the same physical cable (Figure 2–7). The Ethernet terminal would listen to the bus cable, and if it did not see any other terminal transmitting, it could start transmitting a message to any other station. If it sensed a collision from data transmitted simultaneously from another station, both would stop, wait a random time and retransmit. This access protocol is known as CSMA/CD for "carrier sense multiple access, collision detection." Now Ethernet uses unshielded twisted pair wiring and is called 10BaseT. This

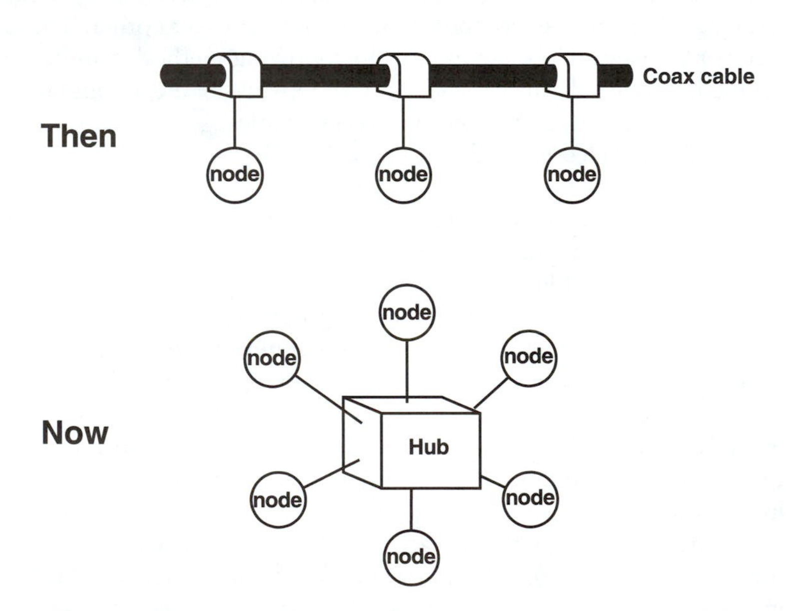


**Figure 2–7.** Ethernet began as a bus network using coaxial cable, but is now primarily a star architecture.

wiring scheme connects each Ethernet station to a "hub" that retransmits to all connected stations, duplicating the function of the original backbone coax cable with simpler, less expensive cabling.

An alternative network connection is on a ring, where all stations are connected to two other stations in the network, forming a ring. Access to the ring is by a "token," a data packet transmitted from one station to the other in sequence. The station with the token can transmit data when it receives permission indicated by the arrival of the token.

The first widely used Token Ring was from IBM, but it has been overshadowed by the popularity of the simpler Ethernet. However, a high-speed Token Ring, called Fiber Distributed Data Interface (FDDI), has become quite popular as a high-speed backbone for large networks (Figure 2–8). FDDI uses two ring connections, one active and one backup, to provide fault tolerance in case of failure of any part of the network.

Whatever type of network architecture is installed, the wiring has become standardized. Ethernet began on a complex, expensive coax cable and Token Ring

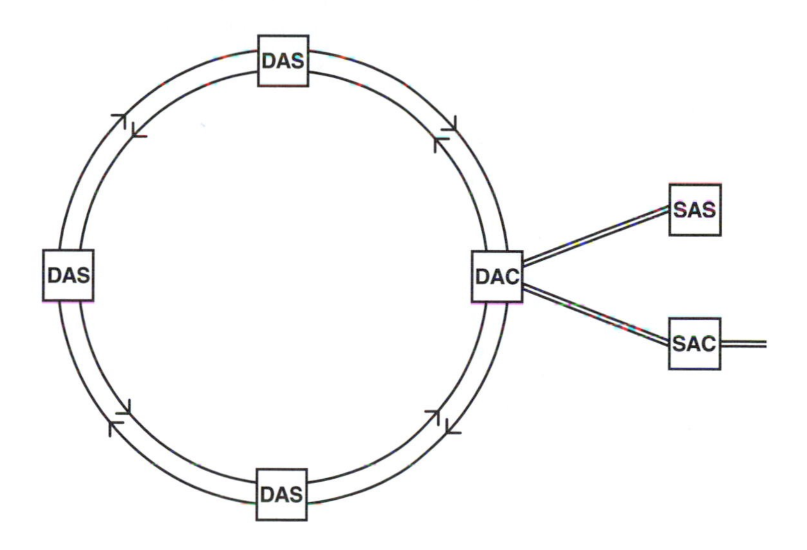


**Figure 2–8.** FDDI uses dual counter-rotating rings to provide fault-tolerant networking.

was on twisted pair cable with individually shielded pairs. Now almost all LANs use unshielded twisted pair (UTP) cables for the desktop connection, and many use optical fiber for the backbone connections (Figure 2–9).

UTP cable has been developed to support most networks up to 100 MB/s (megabits per second). The cable has very tightly controlled physical characteristics, especially the twist on the wire pairs and characteristics of the insulation on the wire, that control the characteristics of the cable at high frequencies.

Normally wire pair cables cannot carry high-bandwidth signals fast enough for LAN communications. The wires also radiate like antennas, so they interfere with other electronic devices. However several developments allowed for the use of simple wire for high-speed signals.

The first development is using twisted pair cables. The two conductors are tightly twisted (one to three twists per inch) to couple the signals into the pair of wires. Each pair of wires is twisted at different rates to minimize cross-coupling. Then, "balanced transmission" is used to minimize electromagnetic emissions (Figure 2–10).

Balanced transmission works by sending equal but opposite signals down each wire. The receiving end looks at each wire and sees a signal of twice the amplitude carried by either wire. While each wire radiates energy because of the signal being transmitted, the wires carry opposite signals so the two wires cancel out the radiated signals and reduce interference.

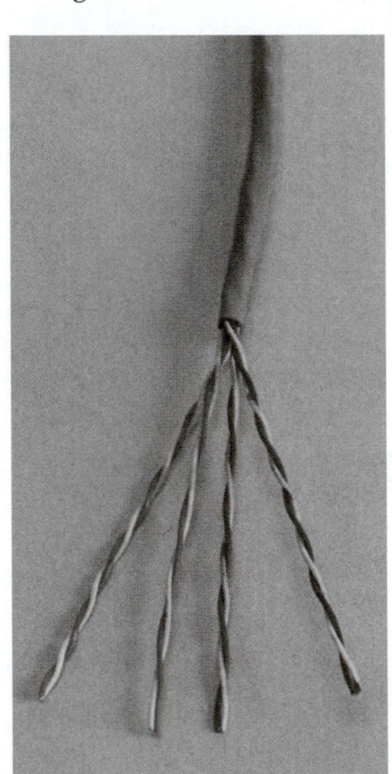

**Figure 2–9.** Unshielded twisted pair (UTP) cable is most the common medium for networking today.

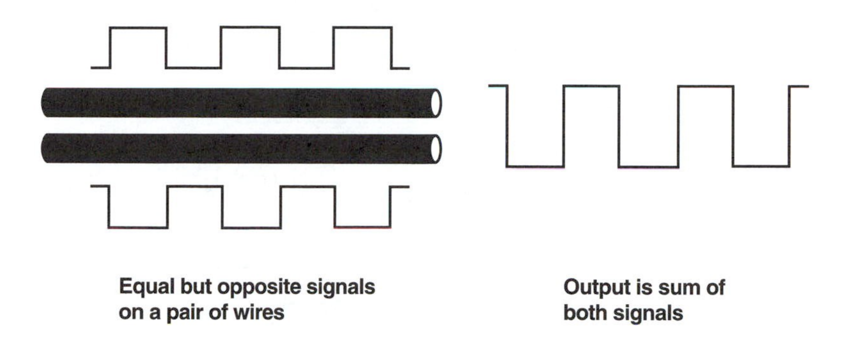


**Figure 2–10.** Balanced pair transmission minimizes emission of radiation from and crosstalk among pairs.

Using these techniques, and having two pair of wires sending signals in opposite directions, it has been possible to adapt unshielded twisted pair (UTP) cables to work with Ethernet and Token Ring networks. At higher speeds, signals are compressed and encoded and multiple pairs are used for transmission in each direction to allow operation with networks of over 100 megabits (MBs) per second.

UTP has four twisted pairs in the cable, two of which are used to simultaneously transmit signals in opposite directions, creating a full duplex link. Some of the higher-speed networks now use all four pairs to reduce the total bandwidth requirement of any single pair.

Some cables have been offered with high performance guaranteed only on two of the four pairs, but these cables can be troublesome if the cable plant is ever upgraded to high-speed networks. Although most networks only use two pairs of the four pairs available, it is not possible to use the other pair for telephone or other LAN connections, because of possible crosstalk problems.

Today, all network wiring has migrated to a "physical star" network with all cables going out from a central communications room. This network topology is specified by the EIA/TIA 568 standards (Figure 2–11), about which you will hear much more later.

These are the basic considerations of network systems. All of the confusing terms associated with networks simply describe things associated with these basics. In reality, the basic operations of networks are fairly simple. The real problem exists in the methods (and it seems like they all have some new name) used to accomplish parts of these basic operations. As we continue through our discussion of networks, do not lose track of the fact that all the other details are merely methods of accomplishing these basics, and nothing more—no matter how terrifying they sound.

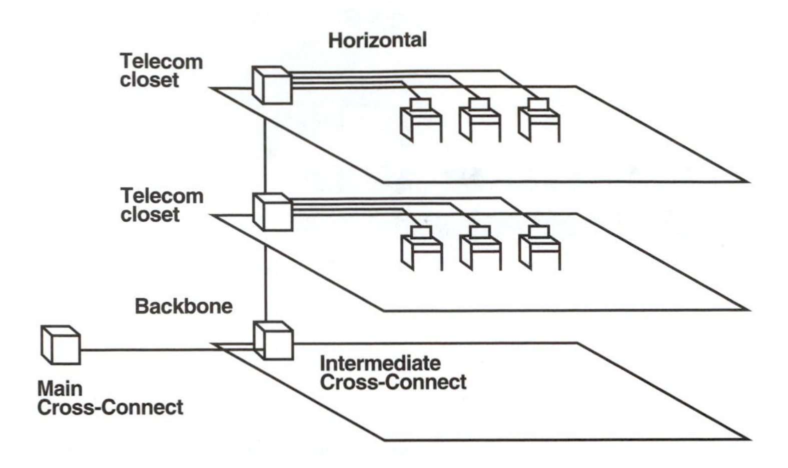


**Figure 2–11.** Network cabling per EIA/TIA 568.

## CHAPTER REVIEW

1. Modern telephones operate on essentially the same principles that were developed 100 years ago. T or F
2. What type of circuit (parallel or series) does a phone operate under?
3. What is the bandwidth of telephone signals?
4. What machine is similar to a black-and-white television?
5. How much bandwidth does a standard TV channel use?
6. What was the first name given to the type of cable capable of transmitting high-speed signals with adequately low loss?
7. List the four parts of a coaxial cable.
8. If the coax cable is not terminated at both ends, what type of problem might you have?
9. In a "tree-and-branch" system, what happens to everyone downstream from a failed amplifier?
10. Using a coax cable capable of carrying a 1 GHz bandwidth, how many channels can you expect in your home?
11. What is the advantage of networking computers?
12. What signals does a digital transmission system transmit? What signals are transmitted by analog system?
13. What type of transmission network is today's TV?

14. What is the vital link between the processor and the network?
    A. Your hands and voice
    B. Pin configuration of the computers
    C. Network interface card
    D. Network interaction class
15. The central computer in a network is usually called the ____________.
16. Name the three most common methods of connecting all the computers in a network.
17. What is the center of a star network?
18. Which is the most common cable used for desktop connections in local area networks (LANs)?
19. What adapter would be needed for connecting hardware from an old cable plant to a newer cable plant using UTP cable?
20. What does Fiber Distributed Data Interface (FDDI) use to provide fault tolerance in case of failure?
21. What type of transmission produces an output as a sum of both signals?
22. Since most networks use two of the four pairs, why is it not possible to use the other two pairs for LAN connection?
23. What network topology is specified by the EIA/TIA 568 standard?

CHAPTER

# 3

# OVERVIEW OF WIRING INSTALLATIONS

Some installers do everything, including designing the Network and all the telecommunications equipment to be installed on it, while others only take someone else's design and install all or part of the cabling. This book will give an overview of the whole effort but will concentrate on the cabling installation.

One caution cannot be overemphasized: If you are not fully confident you know how something should be done, make certain you find out how it should be done correctly before you do it. Good training, working with people who have experience, and phone calls to applications engineers at component manufacturers or distributors can save you from a lot of problems.

## CABLE INSTALLATION

Installing cabling, and its associated work, is by far the biggest part of installing telecommunications systems. We will cover cable installation in some detail. For copper wiring, we will begin with telephone systems, follow with video, and finish with networks, as each application uses different components. Almost all fiber optic projects will use similar components, so they will be covered in one section.

Bear in mind that cabling must be installed according to the requirements of the *National Electrical Code®* and those of your local electrical inspector. In addition, there are accepted voluntary standards and guidelines for installation of voice and data networks that were created by groups of vendors to ensure interoperability of their products, as in EIA/TIA 568. These standards may be called out as references for designing networks. Finally, the customer has the last word, and as the old adage says, "The customer is always right!"

Whether installing voice, data, or video cabling, copper or fiber, there are certain general guidelines to follow in any and every installation that will increase your chances of a successful installation:

1. **Work with a correct design.** Make sure the design and all the drawings follow customer design constraints, *NEC*® rules or other requirements to pass inspection, and other good engineering practices. Watch the cable lengths, as there may be problems with networks if the limits are exceeded.
2. **Choose the proper components.** Make sure the cable is properly rated for the application (e.g., plenum or riser) and marked accordingly. If Cat 5 cable is specified for a LAN, use only Cat 5 rated plugs and jacks, patch panels, and patchcords to maintain Cat 5 performance.
3. **Never abuse cable.** Observe pulling tension and bending radius limits, both during installation and afterward. Riser cables have special limitations that must be followed. Do not overfill conduit or run cable too far unsupported. If you use cable ties, make them snug only, then cut off the tails.
4. **Watch out for interference.** Keep copper signal cables away from power cables, fluorescent lights, electric motors, and other sources of electromagnetic interference.
5. **Consider upgrade requirements.** If the user plans major upgrades in the network someday, consideration of these plans at the outset may save tremendous expense in the future. By installing a higher-rated component, such as Cat 5 cable instead of Cat 3, it may be possible to use the same cabling for future network expansion or speed increases. Fiber optic cables with both multimode and singlemode fibers can be installed at small additional expense and can save installing more cables in the future if new applications call for higher bandwidth.
6. **Think about moves and changes.** Even the EIA/TIA 568 committee is working on a "zone cabling" specification, allowing local drops for ten or so users away from a telecom closet. By consolidating cabling to the zone drop, the cost of installation and the cost of moves and changes can be cut.

## INSTALLING EQUIPMENT

Installing telecommunications equipment is generally fairly simple. It must be properly placed, securely mounted, powered, and properly protected.

### Telephone Equipment

As we cover this material, we will begin with telephone wiring, which is the root technology of all the other communication technologies. (Telegraph technology is actually the real beginning, but for practical purposes, we can consider telephone technology as the root.) We will then go through video, and finish with computer networking, the most complex of these installations.

The most common pieces of *telecom* equipment that you will be mounting are the following:

**Telephones.** The main concern here is that the proper phone is connected in the proper location.

**Punchdown blocks.** It is important that punchdown blocks are mounted securely. If not, connections will be difficult to make, and the block will loosen during the termination process.

**Wall jacks and connection modules.** Here there are two concerns—first, that the conductors are terminated in the proper order, and second, that the jack or module is secure, since it may suffer a fair amount of abuse during its useful life.

**Equipment racks.** Larger systems may require equipment racks. These not only must be secure, but must also be installed plumb and level. They must also be grounded properly.

**Central consoles.** Central consoles must be properly placed and must be installed near a source of electrical power.

### Video Equipment

The most common pieces of *video* equipment that you will be mounting are the following:

**Cameras and mountings.** There are many types of mountings, and every camera needs a mounting bracket. First of all, cameras must be mounted securely. This is especially true because a little bit of wiggle in a camera mounting will give you an image that jumps wildly on the monitor's screen. Pan mountings and remote-controlled mountings require a power feed, and tamper-proof mountings obviously must be extra secure.

**Monitors.** Monitors for security purposes are usually installed in some type of frame or case. As long as you screw them in correctly, you will have no problems.

**Equipment.** Items such as switchers, sequencers, video recorders, motion detectors, and other items need to be securely mounted and properly connected. There are no tricks here; all that is required is good workmanship.

**Antennas, satellite receivers.** These items must be secured to withstand years of wind, rain, snow, and so on. Special concerns are wind resistance and lightning protection (surge protection, and in some cases, strike protection).

### Computer Networking Equipment

The most common pieces of *network* equipment that you will be mounting are the following:

**Personal computers.** Obviously, these machines must be properly placed and properly connected. They must also be properly fed with power and provided with surge protection, and sometimes uninterruptible power sources.

**Computer cards.** Especially when modifying an existing system, you will have to install computer cards in personal computers (PCs). This requires some experience with computers, and just a bit of gentleness. Remember that most computer chips can be permanently damaged by static electricity. Use grounding straps whenever working with PCs.

**Minicomputers and small mainframes.** These units require careful grounding (sometimes isolated grounds) and conditioned power. Sometimes they need specially conditioned environments. Use a lot of care when installing large computers, and make sure that you understand the installation completely before you even start it. Article 645 of the *NEC*® covers many of these installations in detail.

**Hubs, multiplexers, routers.** The main concern with these devices is that they will be protected from damage. The various connections to these devices can be disrupted with very little physical force. Make sure that they will be safe during the installation, and in normal use. Not only should you be concerned about the units being hit, but you should also try to locate them somewhere where they will not be tinkered with by employees or maintenance people.

**Peripheral devices.** This term refers to modems, printers, separate drives, and so on. Here the primary concern is making sure that the devices will be safe from harm and tampering.

**Large central units.** Large central units are essentially dedicated computers connected to switching units. When installing these, you have several concerns. First, you must be sure that there is plenty of space to install the unit and to access it during the installation process and for servicing afterward. You must also be sure that the unit is properly supplied with power and properly grounded. Some large computers have strict guidelines for cooling. Rarely do they need liquid cooling, as did the early mainframes, but they may need air-conditioning for reliable operation.

**Software.** All computer network installations will require installing network operating system software and configuring the network equipment. This is not a job for amateurs! If you have an installation that involves software and network configuration, bring in an expert!

## SAFETY IN INSTALLATIONS

You should never forget about safety in installations. Numerous possibilities for injury exist, and only vigilance will prevent accidents on the job. Keep all these issues in mind when doing installations:

1. **Always wear proper clothing.** Wear a hard hat and safety glasses at all times. Gloves can prevent cuts when working around sheet metal. Even steel-toed shoes may be recommended on construction sites.
2. **Locate and avoid all electrical wiring.** The wire you are installing is a conductor and can cause electrical shock if shorted to an electrical source. Identify any cables that are to be cut and removed before you do any cutting. Know where all electrical cables are and avoid them. (It will be necessary to identify electrical cables to avoid electrical interference in the communication cables anyway.)
3. **Observe ladder safety rules.** Many installations require work above ceilings on ladders. Exercise caution when working on ladders.

## MANAGING INSTALLATIONS

When you begin to take on communications projects, it is critical that you know how to supervise the work of your people. Since we are familiar with power wiring, we think of supervision primarily as making sure that the work gets done in time. Making sure the work is done right is a consideration as well, but it is not the biggest thing in our minds. With telecom work, however, getting it done right is more difficult than getting it done in time.

The problem is this: Mistakes in telecom work (regarding terminations in particular) are difficult to detect. A mistake that could keep the entire system from working might not show up at all until the system is completely installed and turned on. (Think about that for a minute—it is a scary situation.)

So, properly supervising a telecom installation means that you must be able to make sure that your work is good. Yes, it must be done on time; but being on time is meaningless if the work has to be replaced.

Job number one is to ensure that all the terminations are done correctly. This is by far the most important part of supervision. Make sure that your people:

- are properly trained
- have all the right parts
- do not rush

- have a well-lit work area
- have test equipment and use it

Next, be very sure that your people mark every run of cable and termination well. Spend money on cable markers and numbers; spend time on written cable and termination schedules. *Do not* lose track of which cable is which.

Look over the shoulders of your people on the job to make sure they are doing things right. Terminating telecom cables is fine work; make sure that your people work like jewelers, not like framing carpenters.

## INSPECTION

Inspection is a bit of a wild card in telecom installations. Electrical inspectors do not always inspect communication wiring. Nonetheless, take a moment to check with local electrical inspectors before you do any work in their jurisdictions. And obviously you should be very familiar with the low-voltage requirements of the *NEC*®.

In most cases, the inspector of your installation will be the same person who signs your contract, although in some cases, the inspector will be a third party. Make sure you know who will inspect your work before you give your customer a final price. You must know what the inspector will expect of you, and what he or she will be looking for. Be especially careful of third-party inspectors, since they are getting paid to find your mistakes.

The bottom line in installation quality is signal quality and strength from one end of the network to the other. But be careful of other details that may be noticed by the inspector. Among other things, many inspectors will give a lot of attention to proper cable marking, mechanical protection, and workmanship. Pay attention to any detail that the inspector is likely to look for. And be sure to give this a good deal of thought before you begin, or even bid, the project.

## CHAPTER REVIEW

1. What mandatory standard establishes safety standards for installation of cable?
2. What must you keep copper signal cables away from?
3. What are the most common pieces of telecom equipment?
4. What kinds of CCTV equipment might you be mounting?
5. List seven components of computer networks.
6. When wiring telecom projects, what is more important than getting the job done on time and to get paid?
7. What is the bottom line in installation quality?

CHAPTER

# 4

# STRUCTURED CABLING

Several factors have greatly influenced the development of standardized structured cabling systems for building networks. The first event was divestiture. With the original breakup of AT&T, the single-point development of telecommunications standards evaporated, to be replaced by a less powerful entity, Bellcore, and a competitive marketplace. As networks and communications systems became more popular and widespread, vendors realized that users wanted standard cabling systems that would allow using multiple vendors' products and upgrading without complete changes in the cabling.

The model for this structured cabling standard came from the development and use of low-cost UTP (unshielded twisted pair) wiring for computer networks by Ethernet vendors and the development of multivendor standards for UTP cable performance by a cabling distributor, Anixter. As the marketplace and most vendors adopted UTP as the cabling type of choice, it became a de facto standard in the marketplace first and a de jure standard by vendor agreement later.

Today most building communications cabling follows the guidelines developed from this evolution and published by the EIA/TIA TR41.8 committee. Although it is not a mandatory standard, but a voluntary interoperability standard developed by the vendors of the products covered under the standard, it is a common-sense approach to communications cabling that allows simplified interoperability and upgrading.

## STANDARDS FOR NETWORK CABLING

### The Origins of Standards

Widespread usage of any technology depends on the existence of acceptable standards. Users prefer to invest in "standard" solutions to problems, as they promise interoperability and future expandability. Standards must include component standards, network standards, installation standards, standard test methods, and good calibration standards. Standards also include safety, as covered in the *National Electrical Code*®, the only mandatory standard most cable installations must meet.

These standards are developed by a variety of groups working together. Network standards come from Bellcore, ANSI (American National Standards Institute), IEEE (Institute of Electrical and Electronic Engineers), IEC (International Electrotechnical Commission), and other groups worldwide. The component and testing standards come from some of these same groups, plus the EIA/TIA (the merged Electronics Industry Association and Telecommunications Industry Association) in the United States, and internationally from the IEC, ISO (International Standards Organization), and other groups worldwide.

Primary and transfer standards (like "volts" and "meters") are developed by national standards laboratories such as NIST (National Institutes of Standards and Technology, formerly the United States National Bureau of Standards) that exist in almost all countries to regulate all measurement standards. International cooperation is maintained to ensure worldwide conformance to all absolute standards.

We must also discuss "de facto" standards—those generally accepted standards for components and systems that evolved in the marketplace because there were no any de jure standards yet and everyone accepted the work of a supplier.

### De Facto Standards Come First

In any fast-developing technology like computers and communications, there is always resistance for developing standards. Critics say standards stifle technology development. Some critics object because their standards are not the ones being proposed, and in some cases, nobody really knows what standards are best. Under these circumstances, users choose the best solutions for their problems and forge ahead.

The strongest vendors make the de facto standards. For network cabling, Anixter, a large distributor of cabling and network products, developed a set of standards that were adopted by the industry under the EIA/TIA banner, with modifications to suit the many manufacturers participating in the standards process.

### Industry Standards Activities

ANSI coordinates standards activities in the United States. The EIA/TIA, based in Washington, DC, is most active in developing data and voice cabling standards and has the support of all the major vendors of cabling products. EIA/TIA 568 is probably the most quoted standard in communication cabling. It has permeated the cabling industry and developed very high recognition among vendors and users. The TR41.8 committees have developed the following standards relating to building cabling networks:

EIA/TIA 568A Commercial Building Telecommunications Wiring Standard (revision B due soon)
EIA/TIA 569A Telecommunications Wiring Pathways and Spaces
EIA/TIA 570 Light Commercial and Residential Telecommunications Cabling
EIA/TIA 606 Telecommunications Cabling System Administration
EIA/TIA 607 Telecommunications System Grounding and Bonding Requirements

Several technical service bulletins (TSBs) have also been published relating to these standards, to clarify various points in the standards:

TSB-36 UTP Categories 3, 4, and 5 are defined
TSB-40A UTP Connecting hardware for Categories 3, 4, and 5
TSB-53 Additional specifications for STP (shielded twisted pair) hardware
TSB-67 Transmission performance specification for field testing UTP network cabling
TSB-75 Defines "zone distribution systems" for horizontal wiring

Like all standards, 568 is under continuous review. It must be updated to reflect major changes in technology and marketplace conditions. Keeping up with it is important to ensure that the network designer and installer know what is currently acceptable.

### Mandatory or Recommended?

Most standards are not mandatory, that is, they are not required to legally install a network for inspection. In most locales, only *NEC*® (*National Electrical Code*®) for flammability and grounding and FCC rules on the emission of interfering signals are legally required. Many end users, however, will specify network design to EIA/TIA 568A as a common-sense approach to building cabling.

Realize that most of these standards were developed by groups comprised of manufacturers, not users, who cooperated to ensure interoperability of their respective products. No user today wants to be committed to a single vendor for networking products, so interoperability, under the guise of standards, has become expected. Few standards ever reach the point of being mandated by local or federal law.

In spite of all the manufacturers involved in setting these standards, conflicts exist within various standards. For example, 568 is in conflict with the *NEC*® in some areas, which has had to be addressed for future versions of 568.

## BUILDING CABLING ACCORDING TO EIA/TIA 568

The 568 standard has become the model for most premise cabling networks, although it is rarely followed to the letter. It defines a flexible cabling scheme, which allows installing the full cabling network without regard to its ultimate use. Thus a properly installed 568 network will support voice and data (and sometimes even video for short distances) and allow moves, changes, as well as upgrading to different network types to be done more easily. The 568 model defines building cabling according to the model shown in Figure 4–1, a typical building installation.

The electronic equipment is placed in special rooms, with the master equipment for the network in the "main cross-connect." There may be "intermediate cross-connects" located in every building in a large multibuilding network. This equipment is located in an area called the "equipment room" in 568 jargon. The "telecommunications closet" is where the interconnection between the backbone and horizontal cabling systems occurs. It will be located in proximity to the end users (often called the "desktop"). These three locations—the main and intermediate cross-connects, and the telecommunications closet—are where all switches, hubs, and any other networking equipment will be located. All cabling is defined by the necessity to connect all these locations and the desktop of the end user, which is called the "work area."

The 568 standard defines the "backbone" as cabling between telecommunications closets, equipment rooms, and building entrance facilities. Backbone cabling may be UTP, cable, or multimode or singlemode optical fiber.

The model for 568 was AT&T's design guidelines for communications cabling developed from a 1982 survey of seventy-nine businesses located in New York,

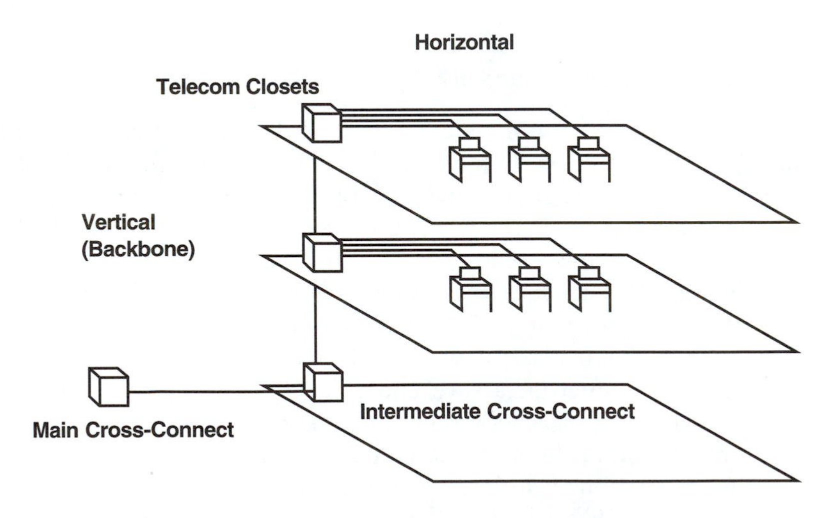


**Figure 4–1** Cabling architecture as per EIA/TIA 568 standards.

California, Florida, and Arkansas involving over ten thousand cable runs. At the time, cabling was used almost exclusively for telephones to wiring closets and PBXs (private branch exchanges or local phone switches), but this usage established a baseline for cable length to be used in 568.

The 1982 AT&T survey determined that 99.9 percent of all stations were less than 300 feet (about 100 meters) from the wiring closet, so that became the goal of the 568 design. With it becoming more difficult to support UTP at that length at gigabit speeds, however, the 1982 survey results showing that about 95 percent of all desktops are within 50 meters of a wiring closet lends credence to establishing a standard for shorter lengths in the near future.

Cabling distances in the current 568 standard for voice and data applications are defined as shown in Table 4–1.

**Table 4–1.** Cabling Used in EIA/TIA 568 and Permitted Distances

| Cable | Distance (m) | Distance (ft) |
|---|---|---|
| 100-ohm UTP cable for data | 90 | 295 |
| 100-ohm UTP cable for voice | 800 | 2,624 |
| Multimode fiber (62.5) | 2,000 | 6,560 |
| singlemode fiber | 3,000 | 9,843 |
| 150-ohm STP (Token Ring) | 90 | 295 |

Note there are several types of cables and applications defined in 568. Besides the normal UTP cable, there is a shielded twisted pair (STP) cable from the cabling system used by IBM Token Ring networks, UTP cable for telephones, and two optical fibers, 62.5/125-micron multimode and singlemode optical fiber. Although all these cable types are offered by the standard, UTP cable is the primary horizontal connection from the telecom closet to the work area and, while it may be used in the backbone also, multimode optical fiber is more common for the backbone connection.

You will also find references in the 568 standard to many other cables for data and voice networks. Many residential telephone applications use a simple 4-conductor cable. Older Ethernet installations may use thicknet (old-style Ethernet coax cable) or thinnet (RG59 coax). Over the years, IBM alone has specified about ten different types of cables for Token Ring, mainframe, terminal, and various other connections. It is beyond the scope of this book to describe all these options, especially since 568 has become so widely accepted. If you have to deal with these other networks, specific directions for them are available from component vendors.

## Components

One of the big advantages of a "structured wiring system," such as that defined by EIA/TIA 568, is the standardization of components. Basically, all connectors, jacks, cables, and punchdowns are identical, making them easier to learn how to use, as well as to manage once they are installed.

### *Cables*

The basic component of the "568" standard is the cable. For most horizontal connections, it will be unshielded twisted pair (UTP) cable. Backbone cabling can be either UTP or fiber optics, with fiber preferred where its higher bandwidth, longer distance capability, or electrical isolation is desired. The fiber optic cable is tightly specified by another EIA/TIA committee, FO2.2, and will be described in a later section of this book.

All UTP cables are comprised of eight 24 AWG solid copper wires, twisted into four pairs, each with a different twist rate. Each pair is color-coded for identification as shown in Figure 4–2. The performance of the cable will fit it into three categories, called Category 3, 4, or 5, depending on its high-frequency capability.

There are cables comprised of twenty five twisted pairs that are used primarily for telephone connections. They are used to reduce the number of cables run between closets and can be more quickly terminated on punchdown blocks. Some manufacturers of these cables are beginning to offer Cat 5 versions of these cables with higher-frequency performance. There are even some 25-pair connectors becoming available. However, most applications still use 4-pair cables due to the possibility of interference of the digital signals running on mixed pairs of cables.

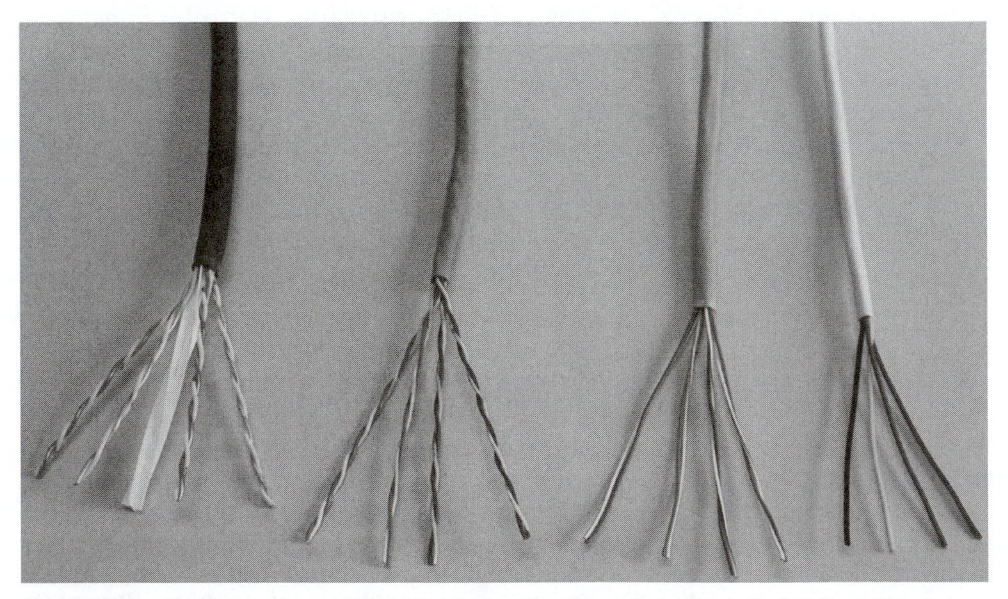

**Figure 4–2.** Structure of unshielded twisted pair cables: from the left, Cat 5e extended, Cat 5, Cat 3, and telephone wire.

**The origins of "category" designations for UTP cable.** Toward the late 1980s, computer networking was growing like mad. Most networks used a proprietary coaxial cable to get adequate bandwidth for the data signals. Manufacturers, however, began to find ways of using enhanced telephone wire, called unshielded twisted pair wire, to send data signals. But every manufacturer built and tested their cable differently, so comparing one cable to another was difficult if not impossible.

With the emergence of UTP cable, one distributor, Anixter, began working with manufacturers, system integrators, and large end users to standardize cables for data applications. In 1989, Anixter developed and published a "Cable Performance Levels" purchasing specification for communications cables.

Under the original Anixter "levels" program, there were three levels of cable performance, allowing customers to select the most cost-effective cables for their application. These three levels were:

Level 1 for "POTS," plain old telephone service
Level 2 for low-speed computer terminal and network (ARCNET) applications
Level 3 for Ethernet and 4/16-MB Token Ring

In 1992, two more levels were added, reflecting newer developments in high-speed networks:

Level 4 for passive 16-MB Token Rings
Level 5 for the copper wire versions of FDDI at 100 MB/s

In the early 1990s, manufacturers turned to the EIA/TIA to help create an interoperability standard for building wiring, based on UTP wire. The EIA/TIA committee adopted the Anixter levels 3, 4, and 5, calling them Categories 3, 4 and 5, and they have since been adopted internationally. Recognizing the need for standards for performance of the components, EIA/TIA participants produced another document, TSB-67, that detailed the testing procedures and specifications that each category must meet.

Of course, in the world of technology, nothing stands still. So today, we have ATM, fast Ethernet, and Gigabit Ethernet to consider. While methods are being used to reduce the bandwidth needs of the network cabling, using encoding schemes like phone modems or multiple pair transmission, it has become obvious that 100 MHz is inadequate for the future of UTP.

**Anixter's proposed updates to "categories" of UTP cables.** There are proposals for higher-performance versions of Cat 5 cable, and even Cat 6 or Cat 7. Again, Anixter is leading the industry into the next stage of defining high-performance UTP cable with its "Levels '97" program. This program is a cooperative effort between Anixter and over forty manufacturers of cabling products. It updates Cat 5 specifications, and adds Cat 6 and Cat 7 UTP specifications as follows:

Level 5 is updated to include delay skew and powersum NEXT testing.
Level 6 expands the frequency range to 155 MHz and tightens powersum NEXT specifications.

Level 7 expands the frequency range even further to 200 MHz and has even tighter powersum NEXT specifications.

Other proposals for higher-bandwidth products include shielded or screened twisted pair designs. You can expect these proposed standards to be hotly debated over the next few years. The upgrading of cable alone is not going to provide higher bandwidth performance. It is necessary to develop interconnection hardware that will provide the bandwidth performance while still being easily field-installed and innovative electronics to compress high-speed signals to fit within the bandwidth limitations of UTP wiring.

At the time of publication, it appears Anixter's proposals for Cat 6 and 7 are much too conservative. Cat 6 may be specified to 250 MHz, and Cat 7 as high as 600 MHz, although it is uncertain if UTP can operate at such high frequencies.

**Cable design and application.** Only Cat 3 and Cat 5 are usually considered for applications today. Cat 4 was an interim specification that has become ignored, since Cat 5 offers tremendously better performance at little increase in price. In fact, in new construction, Cat 5 is often used for everything. Its cost is only slightly higher than Cat 3, and installing one cable simplifies cable management and offers the possibility of future upgrades in performance.

The frequency performance of the cable is mostly determined by the twists in the pairs and the type of insulation used on the wires. The primary specification for each category is the high-frequency limit for its performance (See Table 4–2). The other important specifications are attenuation and cross talk (NEXT for near end cross talk), and for Cat 5, delay skew. All these are specified thoroughly in EIA/TIA 568 and described in more detail here when we discuss testing. Manufacturers that sell Cat 3 or Cat 5 cable follow these specifications carefully in manufacturing their cables.

**Table 4–2.** Frequency Limits of Category-rated Cable

| Cable Bandwidth | |
|---|---|
| **Category** | **Frequency** |
| Cat 3 | 16 MHz |
| Cat 4 | 20 MHz |
| Cat 5 | 100 MHz |

Not all four pairs are used in actual applications. For most LANs, only two pairs are used, one in each direction to allow full duplex, simultaneous bidirectional communications. Due to the limitation on bandwidth and emission of radiation that could potentially affect other electronic devices, the higher-speed networks are migrating toward using all four pairs. (See Table 4–3.)

**Table 4–3.** UTP Cable Pairs Used by Typical Networks

| Network Conductor Use | |
|---|---|
| **Network** | **Pins Used** |
| 10Base-T | 1-2, 3-6 |
| Token Ring | 4-5, 3-6 |
| TP-PMD (FDDI) | 1-2, 7-8 |
| ATM | 1-2, 7-8 |
| 100Base-TX | 1-2, 3-6 |
| 100Base-T4 | 1-2, 3-6, 4-5, 7-8 |
| 100VG-AnyLAN | 1-2, 3-6, 4-5, 7-8 |
| Gigabit Ethernet or 1000Base-T | 1-2, 3-6, 4-5, 7-8 |

In order to cut costs, some manufacturers have used lower-cost insulation materials on two of the pairs, which is permissible for some network applications where only two pairs are used, but the cable will be a problem if it becomes necessary to upgrade to most high-speed networks that use all four pairs.

### *Connections*

On either end of these cables, you will find either a jack or a mating plug, a modular 8-pin design often called an RJ-45 (Figure 4–3). The RJ-45 is a telephone system name, specifically referring to the 8-pin modular design, specific wiring pattern and application, but has become the most common term to describe the connector system used in 568 systems.

Just like the cables, the connectors are available in Cat 3 and Cat 5 versions. If you are installing a Cat 5 network, you *must* use Cat 5 termination hardware throughout. Terminating Cat 5 in Cat 3 hardware will leave you with Cat 3 performance, throwing away the investment in Cat 5 cable.

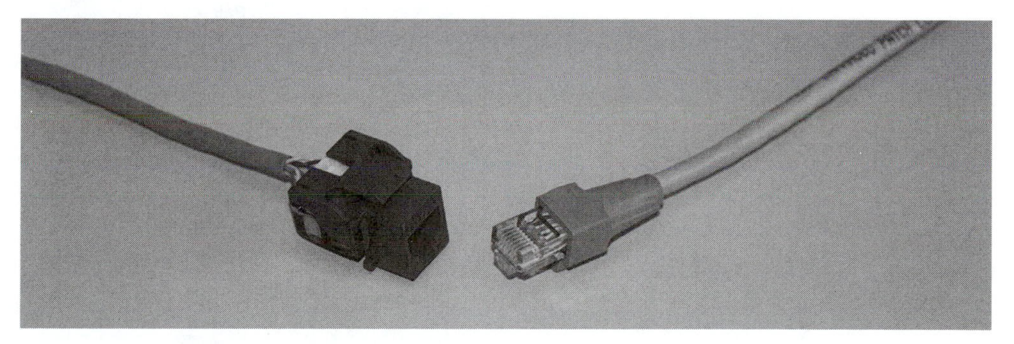

**Figure 4–3.** Standard 8-pin modular plug and jack.

The 568 connector may be wired in several schemes with the four pairs of wire (Figure 4–4). The 568A layout is most widely used, the 568B is an AT&T standard for this connector, and the USOC is a carryover from the phone system, since a USOC jack will mate with a 6-pin modular plug. The view in Figure 4–4 is looking *into* a jack.

The 568A and 568B differ only in one pair being reversed, which may not affect performance, so they are basically similar if used throughout a link. But the two cannot be mixed in a link or you will get a crossing of two pairs that will not make proper connections. Mixing either of these with USOC patterns results in completely messed-up connections.

Some networks use only two pairs, but high-speed networks are beginning to use all four pairs for transmission, as shown in Table 4–3. For any installation, it is vitally important that every pair be properly terminated and checked for proper connections, called wire mapping, even if the cable is currently being used only for POTS.

All 568 terminations are called "insulation displacement connections" or IDC. The connection is made by pushing the wire into a sharp slot in a metal terminal. The terminal slices through the insulation and into the wire, making a good electrical contact. The pierced insulation provides a good seal, preventing degradation of the contact. Modular connectors and punchdown blocks all use this termination method, since it is simple to field-terminate and reliable.

When terminating cables to modular connectors in a Cat 5 network, it is mandatory to maintain the twist in each pair to within 1/2 inch (13 mm) of the connector. The crosstalk of the cable depends on the twists, and untwisting too much wire near the connector will cause the link to fail crosstalk tests.

### *Terminations*

These cables are connected in two ways, with plugs and jacks or with punchdowns. Punchdowns are used as permanent connections in wiring closets, although they can

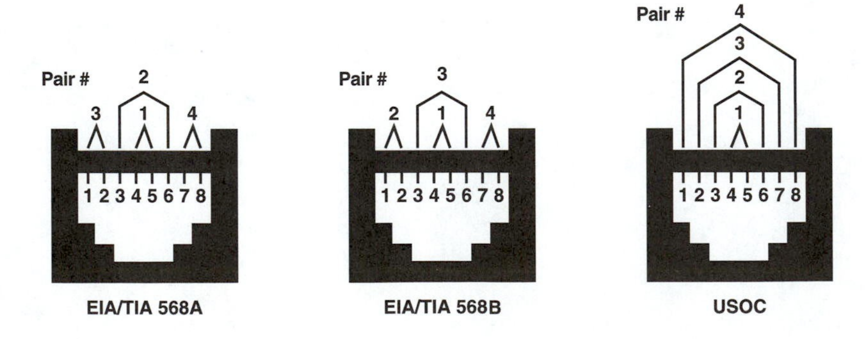


**Figure 4–4.** Pin configuration of modular 8-pin connectors.

be combined with jacks in patch panels to allow flexible patching. If a punchdown or patch panel is used in a Cat 5 network, it must be rated for Cat 5 to maintain the Cat 5 rated performance.

There are several types of punchdowns, including 110, 66, Krone, and BIX. The 110 (Figure 4–5) is the most popular, perhaps because of its flexibility. It can be combined with connecting blocks for mating two cables by punching them into the blocks, or one cable can be punched down and another cable connected with a special patch plug.

The 66 block (Figure 4–6) is the standby in telephone applications, but has only recently been available rated to Cat 5. Unlike the 110, it has rows of four side-by-side pins, with the pair on each side connected. In use the cable is punched down one wire per each pair on the side of the row, then connected with a cross-connect wire punched down on the inside pins of the rows. Alternatively, a bridging clip can be used if wires are connected straight across the rows.

Like connectors, the cable pairs on punchdown blocks must not be allowed to untwist more than 1/2 inch or Cat 5 performance will be lost.

Horizontal and riser distribution cables and patch cables are wired straight through end-to-end so that pin 1 at one end is connected to pin 1 at the other. Crossover patch cables are an exception to this rule. They are wired to cross pairs to allow two Ethernet stations to connect directly without a hub, so the transmitter and receivers must cross in the cable to allow direct connection, as shown in Table 4–4.

**Figure 4–5.** The 100 block is widely used for LAN interconnections, either by (a) punch downs or (b) connectors

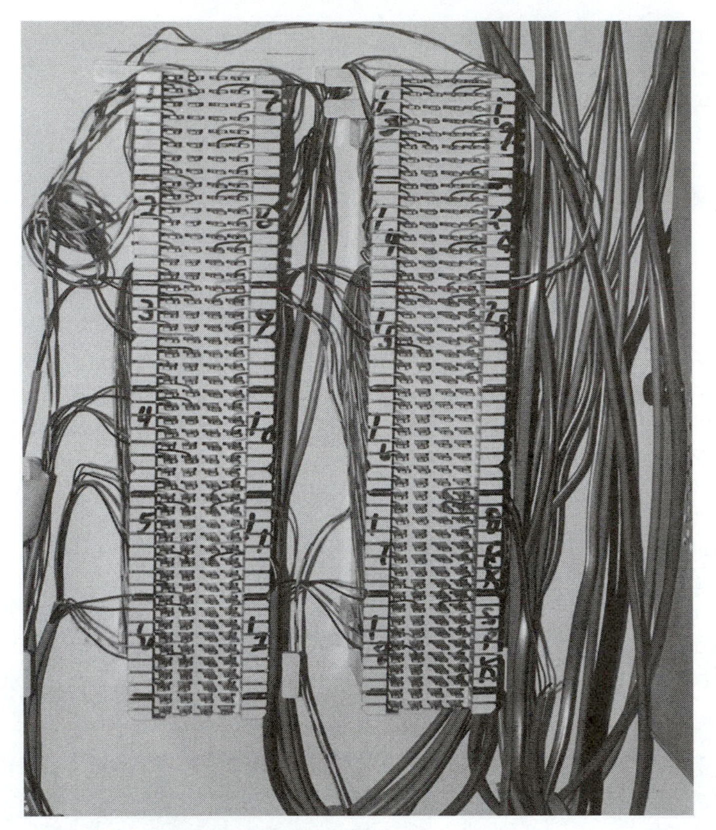

**Figure 4–6** The 66 block is more commonly used for telephone service, but may be used for LANs.

**Table 4–4.** Wiring in a Crossover Patch Cable

| RJ 45 Plug Pin | | | RJ 45 Plug Pin | |
|---|---|---|---|---|
| 1 | Tx+ | → | 3 | Rx+ |
| 2 | Tx– | → | 6 | Rx– |
| 3 | Rx+ | → | 1 | Tx+ |
| 6 | Rx– | → | 2 | Tx– |

Normally, jacks and punchdown blocks are designed so that the installer always punches down the cable pairs in a standard order, from left to right: pair 1 (blue), pair 2 (orange), pair 3 (green), and pair 4 (brown). The white-striped lead is usually punched down first, followed by the solid color. The jack's internal wiring connects each pair to the correct pins, according to the assignment scheme for which the jack is designed—568A, 568B, USOC, or others.

*Patch Panels.* Patch panels offer the most flexibility in a telecom closet (Figure 4–7). All the incoming wires are terminated to the back of the patch panel (again watching the 1/2 inch limit of untwisting pairs). Then patch cables are used to interconnect the cables by simply plugging into the proper jacks.

The termination to the patch panels is usually using 110 style punchdowns on the back of the jacks, although most manufacturers have some proprietary jack termination method. Careful designs of the mounting hardware, including internal twists of the paths to the jacks, maintains Cat 5 performance.

Patch panels can have massive numbers of cables, so managing these cables can be quite a task in itself (Figure 4–8). It is important to keep all cables neatly bundled and labeled so they can be moved when necessary. However, it is also important to maintain the integrity of the cables, preventing kinking or bending in too small a radius which may adversely affect frequency performance.

**Figure 4–7.** Patch panels simplify reconfiguring networks.

**Figure 4–8.** Properly dressed make a neater installation.

### Options in Cable Installations

Structured cabling systems are quite flexible. Within the guidelines of distance and the number of interconnections, one can use combinations of terminations to produce a number of layouts that will support Cat 5 based systems. Figure 4–9 shows some of the options.

Each end that has a jack or patch panel will use patchcords to connect the equipment using the network. The top option (plug to plug) is often used with patchcords to attach equipment in a small office or home environment, where the distances are small and PCs can be connected directly to a hub.

Obviously, the fewer terminations the better. Each termination will cause some degradation of the system. For that reason, the jack to jack or jack to patch panel option is a good choice. "Zone Cabling", the final option shown, is popular with modular offices. A zone box is connected to the TC, perhaps with a 25-pair Cat 5 cable, and local drops to the work area are done with patchcords.

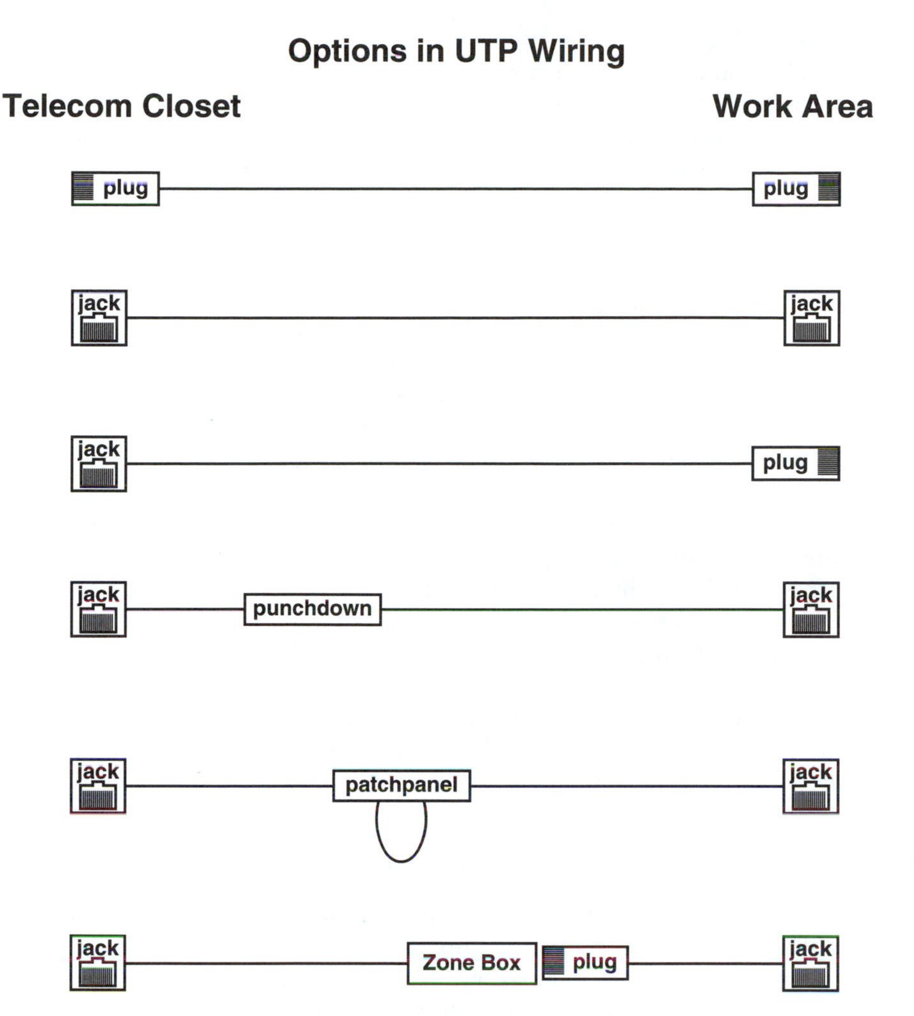


**Figure 4–9.** Many different configurations are acceptable in a Cat 5 rated UTP link.

#### Adapting Hardware to UTP Cabling

The widespread acceptance of UTP cabling has meant that many older designs for coax or STP cables are incompatible with modern cabling. Differences in connector types, balanced versus unbalanced transmission, and cable characteristic impedance means that hardware designed for other applications will normally not operate over this cable. However, adapters called "baluns" (a reference to the BALanced-UNbalanced function) are available to make the conversion. They offer combinations of three distinct functions: converting balanced to unbalanced transmission, matching impedance, and converting connector types.

Baluns are offered to adapt most networks to run on UTP cabling and even to allow hardware designed to run on UTP to use other installed cable types available at a given location. When selecting and using baluns, make certain the proper network operating parameters are specified when obtaining the baluns to get correct ones.

### Future Enhancements

Like everything else that deals with computers and communications, the speed is going up. Cat 5 is marginal to handle newer systems, so there are proposals for Cat 6 and Cat 7 already, with performance specified up to 300 MHz. Already vendors are offering their own versions of high performance cables, but terminating hardware is not yet widely available.

To really use this high performance, connecting hardware may not be enough. Another version of a four pair cable uses a foil shield under the jacket to prevent the transmissions on the cable from causing interference to outside electronics, and to prevent any outside electrical noise from bothering the signals being transmitted on the four pairs inside the cable. This type of cable, called screened twisted pair or ScTP, requires shielded connectors and other hardware to maintain its performance.

New cable and hardware advances seem to be announced daily, so it is important to talk to vendors and read the proper trade journals to keep up to date on the latest technology.

## FIBER OPTICS AND STRUCTURED CABLING SYSTEMS

Fiber optic cabling is listed as an option in the EIA/TIA 568 standard for structured cabling systems. However, it is only listed as an option for Cat 3 or Cat 5 UTP for use up to 100 meters in horizontal cabling, using the same architecture with electronics in every telecom closet. Either multimode or singlemode fiber is specified for use up to 2,000 feet in backbones. The EIA/TIA 568 standard fiber optic connector is the SC type.

While future versions of 568 (568B is due soon) will include more realistic options for fiber, including a "home-run" desktop cabling scheme (Figure 4–10) and zone cabling, fiber is at best an afterthought in 568A. The specified architecture for fiber to the desk is not very cost-effective, and most users and installers prefer the ST connector to the 568SC connector.

A "home run" cable layout connects every desktop through connector panels in local areas, through backbone cables to centralized electronics. This passive distribution scheme is also referred to as "zone cabling," as larger cables are run to a local zone, where smaller cables connect to the individual workstation outlets. Even

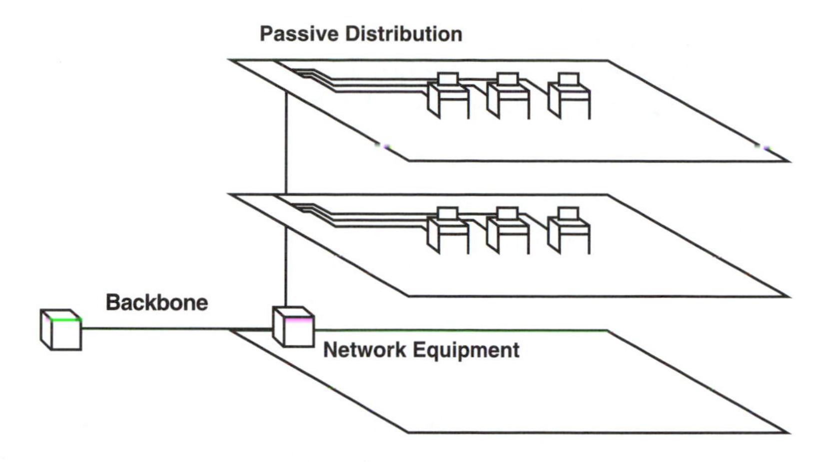


**Figure 4–10.** Optimized premise cabling with fiber optics.

copper cable methods are using this layout on a smaller size, as it simplifies wiring to modular furniture or office clusters.

By installing an optimized fiber optic LAN architecture, with all electronics in one location and longer fiber optic cable runs to the desktop, initial costs will be more competitive with copper wiring. Networking electronics costs less because of the smaller number of electronic interfaces needed, and the management of the system is greatly simplified.

Recognizing that end users are installing fiber in this more reasonable layout, the 568 committee is adding both centralized and zone cabling to the 568B release.

## CHAPTER REVIEW

1. What type of copper cable has become the standard in the LAN marketplace?
2. List the issues that standards *must* address.
3. What is the most quoted standard in communication cabling?
4. What are, in most locales, the mandatory standards for cabling?
5. Who developed the 568 standards (manufacturers or users)?
6. Name the three locations where the switches, hubs, and other networking equipment are located.
7. What three cables are used in the backbone?
8. What was the distance of the UTP wiring with the design of 568, in feet and meters?

9. What is normally run as the connection from the telecom closet to the work area or desktop?
10. What components were made identical under the standardization of components from the EIA/TIA 568?
11. What colors, combined with white, identify the four pairs of UTP cable?
12. Who originally developed cable performance levels?
13. What do the new Level 5, Level 6, and Level 7 cables have in common?
14. What determines the frequency performance of the cable?
15. What is the common mating plug or jack usually called instead of its proper name—modular 8-pin connector?
16. Which pairs of wires are reversed in the 568A and 568B wire scheme?
17. How much can a pair in UTP Cat 5 be untwisted before you fail crosstalk tests?
18. What is the most popular punchdown used for LANs?
19. What type of block is typically used with jacks or a patch panel?
20. What allows older product designs to still be used in UTP-structured cabling systems, and what does this product do?
21. What optical fibers are allowed in 568 structured cabling systems?
22. What fiber optic connector is specified by 568?
23. What is another name for a “home run” cabling system?

PART

# 2

# COPPER WIRING

CHAPTER

# 5

# TELEPHONE WIRING

## RESIDENTIAL OR COMMERCIAL?

You are undoubtedly familiar with residential phone wiring. Most everyone has by now wired a new phone jack themselves, or at least has seen how the phone wiring runs (often chaotically) throughout houses. New construction tends to be much more logical, with most wiring done in a star pattern out from a central location in the house, to facilitate troubleshooting and expansion.

Small businesses, with more lines and phones, often use small modular systems with microprocessor technology to offer services similar to large phone systems, but on a smaller scale. Such installations will mount on a wall in an office or closet area, connecting to a large copper cable for outside line connections and multiple UTP cables to each work area.

The advent of structured cabling systems and general acceptance of EIA/TIA 568 standards for communications wiring in commercial installations have changed the nature of commercial telephone wiring. The simple 2- or 4-conductor wire, plugs or modular outlets, and daisy-chain wiring typical of telephone installations is now common only in residential installations. While this section describes the typical telephone installation, bear in mind that in larger commercial installations, a structured wiring system using UTP and following 568 guidelines is more typical.

This section primarily refers to residential wiring with a section on small office telephone systems. Refer to the chapters on structured cabling systems to understand commercial telephone installations better.

## INSIDE WIRING

Inside wiring is all telephone wire that is inside a telephone company customer's premises and is located on the customer's side of the *network interface* (NI). This wiring comprises the vast majority of telephone wiring.

The network interface (NI) is the physical and electrical boundary between the inside wiring and the telecommunications network. The NI can be any type of telephone company–provided connecting point. Usually it is some type of small box with connectors and/or modular jacks inside. The NI is almost always mounted on an exterior wall, and may be either inside or outside of the structure. For single-family dwellings, the NI is most commonly installed outside, and for commercial or multifamily dwellings, it is most commonly installed indoors in a closet or basement.

A telephone circuit runs from a home or business to the local telephone company switching office. At the local office it is connected to equipment that hooks up to the national telephone network.

Generally, the telephone wiring entry to a structure will be located at or near the same place as the entry for electrical wiring. The NI will be placed near this entry. If an NI is not in place, any existing telephone company–provided modular jack may be used to connect newly installed customer-provided inside wire to existing inside wire.

The NI must be located inside the customer's premises at an accessible point for several reasons:

- Connection through a telephone company–provided modular jack is required by the Federal Communication Commission's (FCC) registration program.
- Utilization of a jack makes it easier to connect or disconnect customer equipment or wire to the telecommunications network.
- Having the jack inside the customer's premises is not required but helps ensure the customer's privacy of communication and helps to prevent unauthorized use.
- Utilization of a jack forms a boundary for the ending of the network service and the beginning of the inside wiring and equipment. This creates a dividing point for the wiring, and for deciding who is responsible for it, should a problem develop.

The point of location for the NI will be determined by the telephone company, although on new construction, you will have to verify the location with phone company representatives prior to beginning the work.

When you complete the wiring, you will plug your wiring directly to the NI or other telephone company–provided modular jack. For residential wiring, the end of your wire should have a modular plug on it to enable you to connect to the NI or telephone company–provided modular jack. For commercial premises, you will more often than not have to punch your connections down at the NI.

For existing installations, connecting points between your inside wire and your telephones may be of several types, depending on when your phones were installed:

- Modular: Most recently installed telephones are connected to the inside wire via a modular system, which, for desk-type phones (Figure 5–1), consists of a miniature plug at the end of the telephone cord and a matching jack on the wall or baseboard.
- Connecting Block: Wall-mounted phones have a pair of slots and a sliding modular plug on the back (Figure 5–2). The phone is attached to a "connecting block" on the wall, which has two rivets that fit into the slots and a modular jack that accepts the plug located on the back of the phone.
- Permanently wired: The telephone is connected directly to the inside wire and cannot be unplugged. The connection point is usually a small, square plastic box near or on the baseboard by the floor.
- Four-prong: On some desk-type telephones, there is a round or rectangular four-prong plug at the end of the telephone cord. The four-prong plug plugs into a jack with four holes. Telephones equipped with such plugs may be plugged in and unplugged easily, enabling you to move them from room to room as needed.

To convert permanently wired phones or those with four-prong plugs to a modular system, you will need an appropriate converter kit, available at most stores that sell telephone accessories.

## INSTALLATION AND CODE REQUIREMENTS

All building and electrical codes applicable in your state to telephone wiring must be complied with. Article 800 of the *NEC*® covers communication circuits, such as telephone systems and outside wiring for fire- and burglar-alarm systems. Generally these circuits must be separated from power circuits and grounded. In addition, all

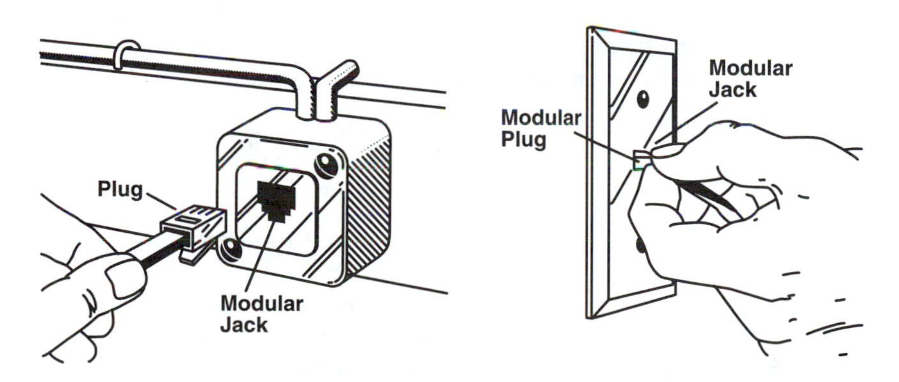


**Figure 5–1.** Typical wall mounted and flush mounted modular phone jacks.

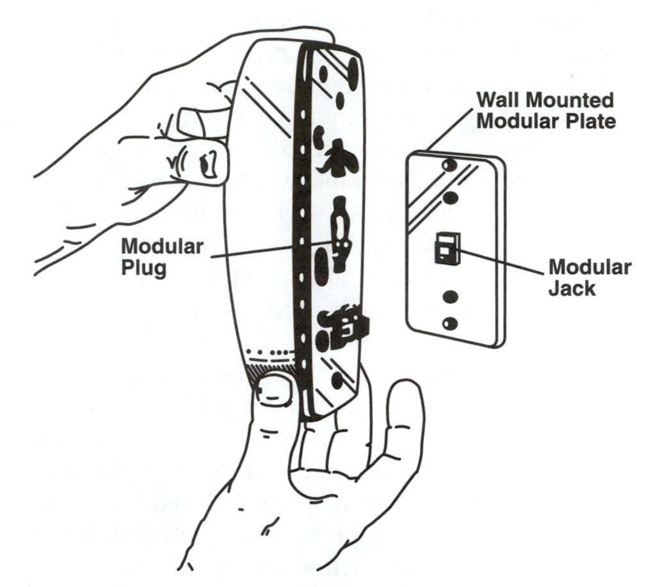


**Figure 5–2.** Modular jack and plate for mounting wall phones.

such circuits that run out of doors (even if only partially) must be provided with circuit protectors (surge or voltage suppressers). The requirements for these installations are as follows.

For conductors entering buildings, if communications and power conductors are supported by the same pole, or run parallel in span, the following conditions must be met:

1. Wherever possible, communications conductors should be located below power conductors.
2. Communications conductors cannot be connected to crossarms.
3. Power service drops must be separated from communications service drops by at least 12 inches.

Above roofs, communications conductors must have the following clearances:

1. Flat roofs: 8 feet.
2. Garages and other auxiliary buildings: None required.
3. Overhangs, where no more than 4 feet of communications cable will run over the area: 18 inches.
4. Where the roof slope is 4 inches rise for every 12 inches horizontally: 3 feet.

Underground communications conductors must be separated from power conductors in manhole or handholes by brick, concrete, or tile partitions. Communications conductors should be kept at least 6 feet away from lightning protection system conductors.

## CIRCUIT PROTECTION

Protectors are surge arresters designed for the specific requirements of communications circuits. They are required for all aerial circuits not confined with a block. ("Block" here means city block.) They must be installed on all circuits within a block that could accidentally contact power circuits over 300 volts to ground. They must also be listed for the type of installation.

Metal sheaths of any communications cables must be grounded or interrupted with an insulating joint as close as practicable to the point where they enter any building (such point of entrance being the place where the communications cable emerges through an exterior wall or concrete floor slab, or from a grounded rigid or intermediate metal conduit).

Grounding conductors for communications circuits must be copper or some other corrosion-resistant material, and they must have insulation suitable for the area in which they are installed. Communications grounding conductors may be no smaller than No. 14. The grounding conductor must be run as directly as possible to the grounding electrode, and it must be protected if necessary.

If the grounding conductor is protected by metal raceway, it must be bonded to the grounding conductor on both ends. Grounding electrodes for communications ground may be any of the following:

1. The grounding electrode of an electrical power system
2. A grounded interior metal piping system (Avoid gas piping systems for obvious reasons.)
3. Metal power service raceway
4. Power service equipment enclosures
5. A separate grounding electrode

If the building being served has no grounding electrode system, the following can be used as a grounding electrode:

1. Any acceptable power system grounding electrode (See *NEC*® section 250-81.)
2. A grounded metal structure
3. A ground rod or pipe at least 5 feet long and 1/2 inch in diameter (This rod should be driven into damp earth, if possible, and kept separate from any lightning protection system grounds or conductors.)

Connections to grounding electrodes must be made with approved means. If the power and communications systems use separate grounding electrodes, they must be bonded together with a No. 6 copper conductor. Other electrodes may be bonded also. This is not required for mobile homes.

For mobile homes, if there is no service equipment or disconnect within 30 feet of the mobile home wall, the communications circuit must have its own grounding electrode. In this case, or if the mobile home is connected with cord-and-plug, the communications circuit protector must be bonded to the mobile home frame or grounding terminal with a copper conductor no smaller than No. 12.

## INTERIOR COMMUNICATIONS CONDUCTORS

Communications conductors must be kept at least 2 inches away from power or Class 1 conductors, unless they are permanently separated from them or unless the power or Class 1 conductors are enclosed in one of the following:

1. Raceway
2. Type AC, MC, UF, NM, or NM cable, or metal-sheathed cable

Communications cables are allowed in the same raceway, box, or cable with any of the following:

1. Class 2 and 3 remote-control, signaling, and power-limited circuits
2. Power-limited fire protective signaling systems
3. Conductive or nonconductive optical fiber cables
4. Community antenna television and radio distribution systems

Communications conductors are not allowed to be in the same raceway or fitting with power or Class 1 circuits, and they are not allowed to be supported by raceways unless the raceway runs directly to the piece of equipment the communications circuit serves.

Openings through fire-resistant floors, walls, and so on must be sealed with an appropriate fire-stopping material.

Any communications cables used in plenums or environmental air-handling spaces must be listed for such use.

Communications and multipurpose cables can be installed in cable trays.

Any communications cables used in risers must be listed for such use.

## TELEPHONE WIRING COMPONENTS

Before beginning any wiring job, you must plan ahead. Determine what types of components you will need. There are several types of standard components associated with telephone wiring.

### Telephone Wire

Conductors in telephone wires shall be solid copper, #22 AWG minimum, and have at least four insulated conductor wires, which may be colored red, green, black, and yellow, or which may follow standard color coding. The conductors shall have an outer plastic coating protecting all conductors with a 1500-volt minimum breakdown rating. Although one phone line only needs two conductors (for "Tip" and "Ring"), the other two conductors are provided for powering dial lighting on some phones or to allow a second phone line to be easily installed. (See Tables 5–1 and 5–2.)

**Table 5–1.** Typical Inside Wire

| Type of Wire | Pair Number | Pair Color Matches | |
|---|---|---|---|
| 2-pair wire | 1 | Green | Red |
| | 2 | Black | Yellow |
| 3-pair wire | 1 | White/Blue | Blue/White |
| | 2 | White/Orange | Orange/White |
| | 3 | White/Green | Green/White |

**Table 5–2.** Inside Wire Connecting Terminations

| Wire Color | | Wire Function | |
|---|---|---|---|
| 2-pair | 3-pair | Service w/o Dial Light | Service w/Dial Light |
| Green | White/Blue | Tip | Tip |
| Red | Blue/White | Ring | Ring |
| Black | White/Orange | Not used (2nd line—Tip) | Transformer |
| Yellow | Orange/White | Ground (2nd line—Ring) | Transformer |

### Bridges or Cross-connects

The purpose of a bridge is to connect two or more sets of telephone wires. Some bridges include a cord with a modular plug on the end, which can serve as an entrance plug in connecting your wire to the telephone company–provided NI or modular jack. Other bridges are designed to be placed at a junction where several telephone wires meet. Proper use of bridges will minimize the amount of wire required for the job.

### Modular Outlets

These are the jacks or connecting blocks into which modular phones are plugged. These jacks are known as RJ-11 for two-wire connections, RJ-14 for four wires (two lines), and RJ-45 which may have four, six, or eight connections. There are two basic types: jacks for desk telephone sets, and jacks for wall telephone sets. In shopping for wiring components, you may find several variations for modular jacks. Some attach to the surface of the baseboard or wall, while others are flush-mounted, requiring a hole in the wall. Some also provide a spring-loaded door to cover the jack opening when nothing is plugged into it. This protects the inside of the jack from dust or dirt, which can damage the electronic contacts. These outlets must meet the Federal Communications Commission's (FCC) registration program requirements.

### Typical Fasteners and Recommended Spacing Distances

Wire staples generally are used to secure the cables to structural surfaces. Other types of fasteners are sometimes used, and should be installed using the spacing guidelines in Table 5–3.

**Table 5–3.** Fastener Spacing Guidelines

| Fasteners | Horizontal | Vertical | From Corner |
|---|---|---|---|
| Wire clamp | 16 inches | 16 inches | 2 inches |
| Staples (wire) | 7.5 inches | 7.5 inches | 2 inches |
| Bridle rings** | 4 feet | | 2 to 8.5 inches* |
| Drive rings** | 4 feet | 8 feet | 2 to 8.5 inches* |

*When changing direction, the fasteners should be spaced to hold the wire at approximately a 45-degree angle.

**To avoid possible injury do not use drive rings below a 6-foot clearance level; instead, use bridle rings.

## SECOND LINE INSTALLATIONS

Residences often have two phone lines installed today, either for the convenience of family members or for the use of a modem or fax machine in a home office. The addition of a second line is simplified if the wiring in the house is already four-wire. Ordering another phone line from the local telephone company will result in a second line added to the NI. If four wires are available in the house, determine that one pair (black/yellow) is not attached to a transformer for powering dial lights (or remove the transformer if it is). Then the black wire attaches to "Tip" and the yellow wire to "Ring" on the second line.

Most modular outlets are wired with all four wires, so if the second pair is connected, an RJ-14 (four-wire) connector (Figure 5–3) can be used to connect two line phones directly to the outlet. If two separate phones, or a phone and modem or fax machine, are desired from the single outlet, a special breakout adapter is available to provide two separate phone lines.

## SMALL OFFICE INSTALLATIONS

Small office installations have a larger number of lines, and probably more telephones, than residential installations. The phone company will bring to the customer a large multiple-pair cable and terminate it on a punchdown block on the customer premises. The phone system usually includes a switch box (KSU) that allows all phones to access outside lines, plus boxes for other functions like voice mail, music on hold, alarm call, and so on. A typical installation is shown in Figure 5–4.

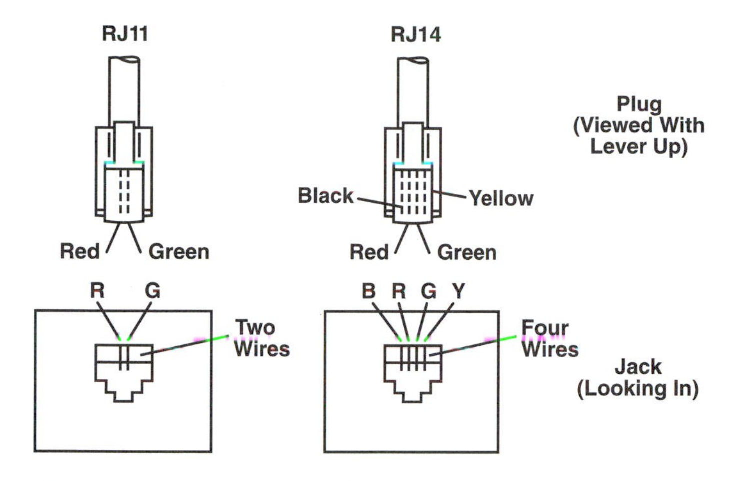


**Figure 5–3.** Wiring for RJ-11 and RJ-14 plugs.

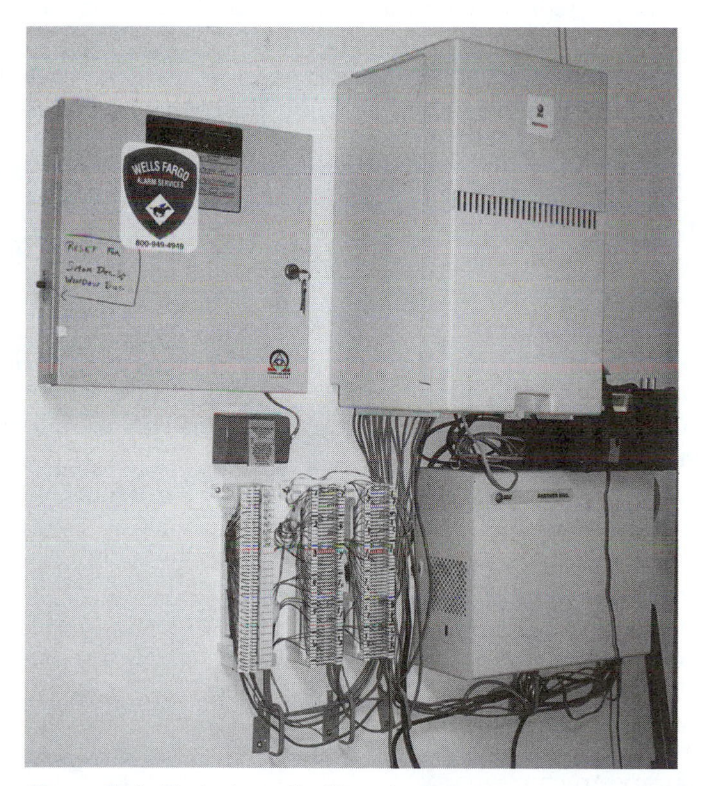


**Figure 5–4.** Typical small office phone system.

The block on the far left is the phone company drop, with a 25-pair cable hidden in the wall and individual line cables terminated in modular plugs connected to the switch (upper right corner). The two 66 blocks are where the switch is connected to the individual phone lines, running out to modular jacks at each work area. By using a terminal block, the lines can be switched easily by changing cross-connect wires.

The box on the middle right is the voice mail option, and the box on the upper left is the alarm system dialer. Note the cables are neatly routed and held in place by guides below the blocks. The phone outlet just below the switch has its cable protected in a stick-on snap-closed raceway, which is used in most areas where wires are not snaked through the walls.

## PLANNING THE INSTALLATION

The general rules for planning and performing a telephone installation are the following:

1. Determine where you want to place the modular outlets. This will likely be determined by the owner of the structure itself, as there are no Code guidelines for telephone wiring.
2. Determine which type of outlet is best for each location. If the jack is likely to be exposed to excessive dust or dirt, use jacks with protective covers.
3. Determine the best path to run the wiring from the NI or other existing telephone company–provided modular jacks to each of the new outlets. Place bridges where two or more paths come together.
4. Inventory the tools you will need to do the wiring job, such as:
   - Screwdriver with insulated handle
   - A pair of diagonal cutters, with insulated grips, to cut wire
   - A tool to strip the wire coating off without damaging any of the four conductors
   - Hammer or staple gun for staples used to attach wire to wall or baseboard
   - Drill, with appropriate-sized bits, to drill holes for screws, anchors, and toggle bolts
   - Key hole saw, if a hole in the wall is necessary, and a drill with a large enough bit to make a hole for the saw blade
5. **Do not** place connections to wiring in outlet or junction boxes containing other electrical wiring.
6. Avoid the following if possible:
   - damp locations
   - locations not easily accessible
   - temporary structures

- wire runs that support lighting, decorative objects, and so on.
- hot locations, such as steam pipes, furnaces, and so on.
- locations that subject wire and cable to abrasion

7. Place telephone wire at least 6 feet from lightning rods and associated wires, and at least 6 inches from other kinds of conductors (e.g., antenna wires, wires from transformers to neon signs, and so on), steam or hot water pipes, and heating ducts.
8. Do not connect an external power source to inside wire or outlets.
9. Do not run conductors between separate buildings.
10. Do not expose conductors to mechanical stress, such as being pinched when doors or windows close on them.
11. Do not place wire where it would allow a person to use the phone while in a bathtub, shower, swimming pool, or other hazardous location. Telephone ringing signals are a shock hazard!
12. Do not try to pull or push wire behind walls when electric wiring is already present in the wall area.
13. Use only bridged connections if it is necessary to establish a splice of two or more wires.
14. Place connecting blocks and jacks high enough to remain moisture-free during normal floor cleaning.
15. Do not attach jacks so that the opening faces upward. This increases the potential for damage from dirt and dust.
16. Wires should run horizontally and vertically in straight lines, and should be kept as short as possible between bridges and other connections.
17. Run exposed wiring along door and window casings, baseboards, trim, and the underside of moldings, so it will not be conspicuous or unsightly.
18. Wood surfaces are better for fastening wire and attaching connecting blocks, jacks, and bridges. When attaching hardware to walls, place fasteners in studs (wooden beams behind the walls) whenever possible.
19. If drilling through walls, floors, or ceilings, be careful to avoid contacts with concealed hazards, such as electric wiring, gas pipes, steam or hot water pipes, and so on.
20. If installing cables next to grating, metal grillwork, and so on, use a wire guard or other protective barrier to resist abrasion.
21. Always fasten cables to cement or cinder blocks with screw anchors, drive anchors, or masonry fasteners.
22. Avoid running cables outside whenever possible. If exterior wiring is necessary, drill holes through wooden window or door frames and slope entrance holes upward from the outside. Try to use rear and side walls so the wire will not be as noticeable; place horizontal runs out of reach of children; and avoid placing wiring in front of signs, doors, windows, fire escapes, "drop wires," and across flat roofs.

23. When fastening wire to metal siding, the type of fastener used depends on the type of siding and the method used to install it. Extra caution should be used when working on mobile homes. Mobile homes should be properly grounded. Line voltages present an extreme danger when working on metal. Therefore, proper grounding is very important.

### Small Office Installations

In a small office with several lines, the phone company will usually bring a 25-pair cable into the building and terminate it on a punchdown block. The lines will be tested and marked on the block or nearby. The connection to the telephone equipment will involve punching down wires to connect from the block to the equipment. If the installation involves a small switch that allows all phones to share the lines, the switch will connect every phone in the system to the equipment. Most of these systems now use 4-pair UTP (Cat 3 or 5) cable and 8-pin modular outlets, just like structured cabling systems.

Each phone will connect to a modular outlet connected back to the switching equipment. Many small offices now install networking cabling along with telephone cables to support both services with one installation. For installations like this, refer to the sections of this book on structured cabling.

See Table 5–4 for the standard color coding used for 25-pair cables. Note pairs are coded for pair number and "Tip" and "Ring," which is important for correct installation.

## SEPARATION AND PHYSICAL PROTECTION FOR PREMISES WIRING

Table 5–4 applies only to telephone wiring from the network interface or other telephone company–provided modular jacks to telephone equipment. Minimum separations between telephone wiring, whether located inside or attached to the outside of buildings, and other types of wiring involved are shown in Table 5–5. Separations apply to crossing and to parallel runs (minimum separations).

## INSTALLATION SAFETY

Telephone connections may have varying amounts of voltage in the bare wires and terminal screws. Therefore, before you begin an installation, make sure the entrance point of any existing wire is unplugged from the NI or telephone company–provided modular jack while you are working. This will disconnect any wiring from the telephone network. If you are just connecting a new modular outlet to existing wiring that you cannot disconnect, take the handset of one telephone off the hook. This will prevent the phone from ringing and reduce the possibility of electrical

**Table 5–4.** 25-pair Backbone Cable Color Code

| Pair | Tip<br>Base Color/Stripe | Ring<br>Base Color/Stripe |
|---|---|---|
| 1 | White/Blue | Blue/White |
| 2 | White/Orange | Orange/White |
| 3 | White/Green | Green/White |
| 4 | White/Brown | Brown/White |
| 5 | White/Slate | Slate/White |
| 6 | Red/Blue | Blue/Red |
| 7 | Red/Orange | Orange/Red |
| 8 | Red/Green | Green/Red |
| 9 | Red/Brown | Brown/Red |
| 10 | Red/Slate | Slate/Red |
| 11 | Black/Blue | Blue/Black |
| 12 | Black/Orange | Orange/Black |
| 13 | Black/Green | Green/Black |
| 14 | Black/Brown | Brown/Black |
| 15 | Black/Slate | Slate/Black |
| 16 | Yellow/Blue | Blue/Yellow |
| 17 | Yellow/Orange | Orange/Yellow |
| 18 | Yellow/Green | Green/Yellow |
| 19 | Yellow/Brown | Brown/Yellow |
| 20 | Yellow/Slate | Slate/Yellow |
| 21 | Violet/Blue | Blue/Violet |
| 22 | Violet/Orange | Orange/Violet |
| 23 | Violet/Green | Green/Violet |
| 24 | Violet/Brown | Brown/Violet |
| 25 | Violet/Slate | Slate/Violet |

shock. Disregard messages or tones coming from the handset signaling you to hang up. In addition:

- Use a screwdriver with an insulated handle.
- Do not touch screw terminals or bare conductors with your hands.
- Do not work on telephone wiring while a thunderstorm is in the vicinity.

## INSTALLATION STEPS

1. Install a bridge or some other component to act as an entrance plug for your wire. This plug will connect to the NI or telephone company–provided modular jack. The bridge should have a modular-type cord with a

**Table 5–5.** Separation of Phone Cable from Other Types of Cabling

| Types of Wire Involved | | Minimum Separation | Wire Crossing Alternatives |
|---|---|---|---|
| Electric supply | Bare light or power wire of any voltage | 5 feet | None |
| | Open wiring not over 300 volts | 2 inches | Note 1 |
| | Wire in conduit or in armored or nonmetallic-sheathed cable or power ground wires | None | NA |
| Radio and TV | Antenna lead-in or ground wires | 4 inches | Note 1 |
| Signal or control | Open wiring or wires in conduit or cable | None | NA |
| Communication | CATV coax with grounded shielding | None | NA |
| Telephone drop | Using fused protectors | 2 inches | Note 1 |
| | Using fuseless protectors or where there is no protector wiring from transformer | None | NA |
| Sign | Neon signs and associated wiring from transformer | 6 inches | None |
| Lightning systems | Lightning rods and wires | 6 inches | See wiring separations |

Note 1: If minimum separations cannot be obtained, additional protection of a plastic tube, wire guard, or two layers of vinyl tape extending 2 inches beyond each side of object being crossed must be provided.

plug at the end to insert into the NI or modular jack. Another acceptable type of entrance plug is a length of telephone wire with a modular plug on the end. Do not insert the entrance plug into the NI or modular jack until your wiring is completed.

2. Install all modular jacks in or on walls or baseboards. Use wood screws on wooden surfaces. Drill holes slightly smaller than the diameter of the screws being used to make installation easier. To fasten components to plasterboard walls, use screw anchors or toggle bolts.

3. Run wire to each modular jack, stapling it to the wall or baseboard about every 8 inches. Be sure you do not pierce or pinch the wire with staples. Allow enough wire to make the electric connections to the modular jack attached to the wall or baseboard. (In new installations, the wiring will be in place before the walls are completed.)
4. Strip the plastic coating on the phone wire as needed and connect the colored conductors to the terminals for each modular jack. Trim excess wire and attach the modular jack cover (if any) to the base.
5. When finished, place the plug on the end of your bridge into the NI or telephone company–provided jack.

## TESTING

After installing the wiring, the first step in testing it is to lift the handset of a phone plugged into one of the new outlets, listen for dial tone, then dial any single number other than "0." Listen. If you hear a lot of excessive noise, or if the dial tone cannot be interrupted, you have a problem. Attempt to locate it by using the following "troubleshooting" guidelines. If you cannot locate or repair the trouble yourself, disconnect the defective wiring until you can get the problem repaired.

### Troubleshooting Guidelines

If testing indicates problems in the wiring you have installed, or if problems develop with the phone service later, try to determine if the problems are being caused by your own wire and equipment or by the telephone line. Here are some of the things you can do to try to identify the nature of the problem:

1. Unplug the wire you installed from the NI or telephone company–provided modular jack. Plug any phone (other than the one used when you detected the problem) directly into either of these jacks. If the problem persists, the telephone company lines or equipment may be faulty and you should proceed to step 2; otherwise, see step 3.
2. Dial the telephone company's repair service bureau listed in your directory. Describe the problem you are experiencing; be sure to state that you have installed your own wiring.
3. If the problem no longer exists when you plug another phone into the NI or telephone company–provided modular jack, it probably is being caused by your wiring or equipment. You may be able to localize the source of the problem by plugging the working phone into different outlets and testing each separately as before. Among the possible sources of trouble are broken wires, worn insulation, incorrect (e.g., red and green conductors reversed) or loose connections, and staples put through the wire.

4. *Note:* If you have Touch-Tone® Service and, after lifting the handset of a phone plugged into the new outlet you installed, you hear the dial tone but the Touch-Tone® dial does not operate, unplug the wire from the NI or other telephone company–provided modular jack, reverse the red and green conductors at the outlet, then plug it back into the NI and check the phone again (reversing the polarity). If you still cannot locate the problem, call the telephone company's repair service bureau.

## INSTALLATION CHECKLIST

1. Be sure the entrance plug is unplugged from the NI or telephone company–provided modular jack.
2. Attach each component securely to the wall or baseboard.
3. Run wire to each component, allowing enough extra wire to make electric connections.
4. Make electric connections and put covers on components.
5. Plug the entrance plug into the NI or telephone company–provided modular jack.
6. Plug in telephones and test (see "Testing" instructions).
7. See "Troubleshooting Guidelines" if problems occur.

## CHAPTER REVIEW

1. What is network interface (NI)?
2. The point of location of the NI is always inside the residence. T or F
3. What type of plugs does the telephone system use?
4. What must you have if you have any circuits outside?
5. How far apart should power service drops be from communication service drops?
6. How much clearance must communication conductors have in a roof with a 4/12 slope (4-inch rise and 12 inches horizontally)?
7. Surge arrestors must be put on all circuits with a block that may accidentally contact power circuits over ____ volts.
8. What must ground conductors be made from?
9. What wire gauge for copper is the minimum used in telephone wires?
10. What colors are the four insulated wires normally used in telephone cables?
11. What is the service of a bridge?
12. How far apart should you staple the wire in the vertical or horizontal direction?
13. What angle is suggested when changing direction?

14. What functions other than telephone use may a small office phone system have?
15. Where should you *not* place connections to telephone wiring?
16. Why could the telephone be a problem in the shower?
17. What should one be careful of when drilling into a wall, floor, or ceiling?
18. What should one make sure of when working on mobile homes?
19. What can one do if minimum separations cannot be obtained?
20. List four sources of problems one might encounter in telephone installations.

CHAPTER

# 6

# VIDEO SYSTEM INSTALLATIONS

Cabling for video applications covers two distinct uses: CCTV or closed-circuit TV for security surveillance, and CATV, which stands for community antenna TV (not cable TV) and is used to distribute television signals. Both applications use similar cabling and installation techniques.

## VIDEO CABLING

The high bandwidth of video signals requires the performance of coaxial cables for transmission. Coaxial cable has always been the industry standard for high-bandwidth applications, although it is being replaced by fiber optics for long distance or higher-bandwidth applications. Microwave transmission is used only where required, as its cost is significantly more than the other two methods.

Using a high-quality coaxial cable is essential. For CCTV installations of less than 1,000 feet, RG-59U cable is fine. But for distances of 1,000 to 2,000 feet, RG-11U should be used. Installations of more than 2,000 feet in length require the use of amplifiers to keep the signals at usable levels, or the use of optical fiber as the communication media. Most CCTV installations will use BNC connectors, a bayonet mount connector, or occasionally an "F" connector like CATV.

CATV uses large RG-8 for its trunk lines and the smaller RG-59 or RG-6 for drops to the home and within it. Larger buildings with longer cable runs will use Series 7 or RG-11 cable for its lower loss. The standard connector for CATV is the F connector.

Perhaps most important with CATV cables is proper cable selection and termination. CATV cables are directly connected to the public CATV network. FCC rules limit signal leakage, so it is important to use good cable with proper shielding and terminate properly to prevent signals interference with other electronic devices. In addition, poor termination can cause reflections in the cable that affect the return path, or connection back to the system.

Since more networks now use or plan for cable modems for Internet connections, proper CATV installation becomes more important. A poorly installed installation in a building can cause trouble for thousands of CATV subscribers. The Society of Cable Television Engineers (SCTE) has published standards for the installation of cabling for CATV systems.

## INSTALLATION

Installing video cabling is relatively simple. Coax cables must be installed with care; that is, they may not be pulled beyond their tension limits, and may not be sharply bent. In addition, they must remain safe from physical damage and from environmental hazards. Care must also be taken when strapping or (especially) stapling cables to structural surfaces (walls, ceilings, and so on). If the staple or strap is cinched too tightly, it will deform the cable and alter its transmission characteristics. If the staple or strap is overly tight, the system may not work properly, or even at all.

As mentioned earlier, surveillance cameras must be firmly mounted. In addition to standard wall-mounting brackets, cameras may also be mounted in vandal-resistant cases, which are recommended in trouble-prone areas. Mountings are also available with built-in panning and tilting mechanisms. Obviously, these will cost a bit more than the standard mountings but will provide an additional benefit. However, they cannot be used in all installations, particularly not with video motion detector devices.

## TERMINATION

Coaxial cable connections (Figure 6–1) include a center contact and an outer shell that connects to the shield on the cable. CATV systems use F connectors, a screw thread connector, while CATV may also use BNC, bayonet mount connectors. The N connector is also used on larger coax cables. Note the F connector uses the center conductor of the cable as the center contact, minimizing installation time and cost.

Coaxial connectors are so simple to install that we all can and do install them all the time—for connecting our VCRs—using simple screw-on F connectors. For building wiring with coax, especially for CATV, we need to be more careful to ensure high performance and low signal leakage. Crimp connectors are preferred, and cable should be top-quality. When making up a coaxial connector, the impor-

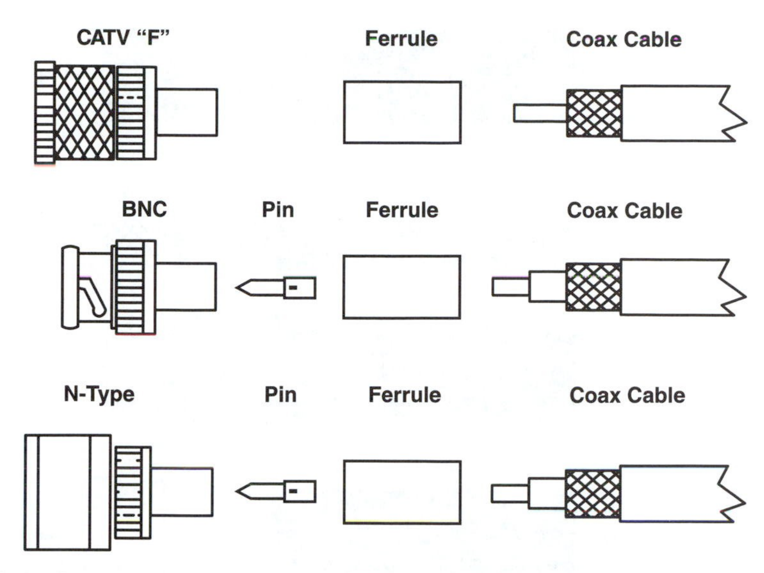


**Figure 6–1.** Common coaxial cable connectors.

tant things are to make sure that the connector type matches the cable type and to make sure the connection is made up securely.

The process of terminating coaxial cables, shown in Figure 6–2, is as follows:

1. The outer insulating jacket must be stripped away, exposing the braided shield.
2. The braided shield must be pulled back, over the outer jacket, leaving the inner insulation and its foil shield exposed.
3. The end of the inner insulation should then be removed, exposing the center conductor.
4. The connector is then placed on the end of the cable and crimped or tightened down. (There are any number of different connectors available, with slightly different termination styles. You must follow the directions for the given connector precisely.)

## CODE REQUIREMENTS

Cable television and security monitoring circuits are covered by article 820 of the *NEC*®. As we all know, the use of such circuits has expanded dramatically in recent years, and is likely to expand much further as the "information highway" continues to develop. The services that will probably do more to promote high-tech telecommunication services than any other are Internet connections by "cable modems" and "video on demand."

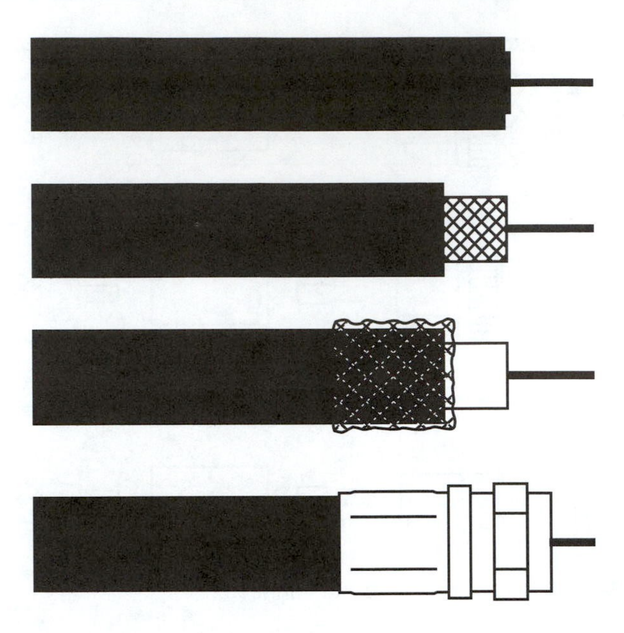

**Figure 6–2.** Termination procedure for coaxial cable.

You will notice that the title of article 820 is "Community Antenna Television." Since very few of us have ever done a community antenna installation, and since we are relating this article to cable television and security monitoring, the following provides a brief explanation of CATV.

Community antenna television (CATV) started decades ago as a means of providing television signals to communities that could not receive broadcast stations, either because of distance or shadow areas where the signal was too weak. Community antennas were installed at remote locations (such as on top of a nearby hill), and signals from them were fed to the homes in the area.

Later, when television signals transmitted via satellite became common, cable TV systems were able to provide a much wider variety of programming than was available via broadcast. Once programming from all over the world was available, the demand for the services became enormous, and cable television companies began to provide services. The new services were based on the same standards and methods as the community antenna systems from which they evolved. Today, cable television has developed into a huge system that serves at least 40 million homes in the United States; and it is growing steadily.

Security monitoring has been used ever since it became possible. At first, when the technology was new and costs were relatively high, security monitoring was used only for more vital uses. Later, as costs came down, its use became widespread.

Article 820 is therefore very broadly defined, covering all radio frequency signals sent through coaxial cables. While the term *radio frequency* (often abbreviated as *RF*) is not defined in the *NEC*®, it would generally include every frequency from several kilohertz to hundreds of megahertz. This would include all types of radio signals, television signals, and computer network signals.

Article 820 does not, however, cover television cabling that is not coaxial (see Fine Print Note [FPN] to section 820-1); it applies to coaxial cable only.

The primary safety requirements of this article are first of all that the voltage applied to coaxial cables cannot exceed 60 volts, and the power source must be energy-limited. Energy limitation is defined by section 725-31.

### Grounding

Grounding is mentioned in both parts B (Protection) and C (Grounding) of article 820. Grounding is particularly required for coaxial cables run outside of buildings. While the *NEC*® does not specifically state that all outdoor cables must be grounded, almost all outdoor runs must be grounded to meet the requirements of section 820-33. The concern with outdoor cable runs is that they will be exposed to lightning strikes. And in addition to direct strikes, outdoor runs of conductors can have substantial voltages induced into the conductors from nearby lightning strikes.

Another concern for grounding is accidental contact with power conductors. Whenever an outdoor coaxial cable could accidentally come into contact with power conductors operating at over 300 volts, grounding is required.

It is always the outer conductor of the coaxial cable that is grounded. The rules are generally as follows:

1. The cable must be grounded as close to the cable's entrance to the building as possible.
2. The grounding conductor must be insulated.
3. The grounding conductor must be at least #14 copper, and must have an ampacity at least equal to that of the coaxial sheath.
4. The grounding conductor must be run in a more or less straight line to the grounding electrode. The run must be protected if subjected to damage.
5. The grounding electrode can be any suitable type. Section 820-40 specifically mentions the following:
   The building grounding electrode
   Water piping
   Service raceway
   A service enclosure
   The grounding electrode conductor, or its metal enclosure

The connection to the grounding electrode can be with any suitable means, as detailed in section 250-115.

If the coaxial cable is grounded to an electrode other than the building electrode, the two electrodes must be bonded together with a #6 or larger copper conductor.

The grounding connection may not be made to any lightning protection conductor, whether it is a grounded conductor or not. You may, however, bond the coaxial grounding electrode to a lightning protection grounding conductor.

Pieces of unpowered metallic equipment such as amplifiers and splitters that are connected to the outer conductor of grounded coaxial cables are considered to be grounded.

### Surge Suppressors

Although not required by the Code, surge suppressors are a practical necessity for virtually all outdoor runs of coaxial cable. Since the pieces of equipment connected to these cables are very sensitive, they are easily damaged by voltage spikes. The most commonly used type of surge suppressor for communications circuits is the metal oxide varistor (MOV). These suppressors are usually called *protectors* in the communications industry. These devices, which are made of sintered zinc oxide particles pressed into a wafer and equipped with connecting leads or terminals, have a more gradual clamping action (clamping is the act of connecting a conductor to ground when the voltage rises too high) than either spark-gap arresters or gas tubes. As the surge voltages increase, these devices conduct more heavily and provide clamping action. And unlike spark-gap arresters and gas tubes, these protectors absorb energy during surge conditions. They also tend to wear out over time.

### Routing of Outdoor Circuits

The Code's rules for the routing of coaxial communications circuits are essentially the same as those for other communications circuits. Specifically, coaxial cables must:

- Be run below power conductors on poles.
- Remain separated from power conductors at the attachment point to a building.
- Have a vertical clearance of 8 feet above roofs (there are several exceptions)
- If run on the outside of buildings, be kept at least 4 inches from power cables (but not conduits)
- Be installed so that they do not interfere with other communications circuits, which generally means that they must be kept far enough away from the other circuits.
- Be kept at least 6 feet from all lightning protection conductors, except if such spacing is very impractical.

### Messenger Cables

The vast majority of outdoor runs of coaxial cables are run with messenger cables (or are specially designed to be messenger cable assemblies themselves). In article 820, the only requirement made for such runs is that the runs be attached to a messenger cable that is acceptable for the purpose, and has enough strength for the load to which it will be subjected (such as the weight of ice or snow, or wind tensions). Messenger cables are covered in more detail in article 321 of the *NEC®*.

### Indoor Circuits

Coax cables indoors are subject to the following requirements:

- They must be kept away from power or Class 1 circuits, unless the circuits are in a raceway, metal-sheathed cables, or UF cables.
- Coaxial cables can be run in the same raceway (or enclosure) with Class 2 or 3 circuits, power-limited fire protective signaling circuits, communications circuits, or optical cables.
- They may not be run in the same raceway or enclosure with Class 1 or power conductors. Exceptions are made if there are permanent dividers in the raceway or enclosure, or in junction boxes used solely as power feeds to the cables.
- They are allowed to be run in the same shaft as power and Class 1 conductors, but in these cases, they must remain at least 2 inches away. (But, as noted earlier, this applies to open conductors, not to conductors in raceways, metal-sheathed cables, or UF cables.)

### Cable Types

The *NEC®* goes into great detail on designated cable types. Many of these requirements apply more to the cable manufacturer than to the installer. Nonetheless, the proper cable type must be used for the installation. The *NEC®* designations and their uses are as follows:

- **Type CATVP.** CATVP is plenum cable (hence the "P" designation), and may be used in plenums, ducts, or other spaces for environmental air.
- **Type CATVR.** CATVR is riser cable, and is suitable (extremely fire-resistant) for run installation in shafts or from floor to floor in buildings.
- **Type CATV.** CATV is general-use cable. It can be used in almost any location, except for risers and plenums.
- **Type CATVX.** CATVX is a limited-use cable, and is allowed only in

dwellings and in raceways.

You will find in practice that many coax cables are multiple-rated. In other words, their jacket is tested and suitable for several different applications. In such cases, they will be stamped with all of the applicable markings, such as both CATV and CATVR. (Notice that all the cable types start with "CATV," which refers to community antenna television, not *cable television.*)

Trade designations generally refer to the cable's electrical characteristics; specifically, the impedance of the cable. This is why different cable types (RG59U, RG58U, and so on) should not be mixed. Even though they appear to be virtually identical they have differing levels of impedance, and mixing them may degrade system performance.

### Substitutions

The Code defines a clear hierarchy for cable substitutions. The highest of the cable types is plenum cable (CATVP); it can be used anywhere at all. The next highest is riser cable (CATVR); it can be used anywhere, except in plenums. Third on the list is CATV, which can be used anywhere except in plenums or risers. Last is CATVX, which can be used only in dwellings and in raceways.

The 1993 Code also permits multipurpose cables to be used for CATV work. But, as always, only plenum types of cables can be used in plenums, riser types in risers, and so on. The formal listing is shown in Table 820-53 in the *NEC*®.

## CHAPTER REVIEW

1. What do the acronyms CCTV and CATV stand for?
2. What type of copper cable is used for the high-bandwidth CATV and CCTV signals?
3. What needs to be done to install cables more than 2,000 feet?
4. What cables and connector does CATV use?
5. How does one prevent signal interference with other electrical devices?
6. What can alter a cable's transmission?
7. What type of mounting cannot be used with built-in motion detector devices?
8. Why was CATV created?
9. What made CATV grow to the point of providing the service to over half the homes in the United States?
10. *NEC*® article 820 refers to what type of cable?
11. Why is grounding important to outdoor cables?
12. What gauge conductor is used for outdoor grounding?
13. To protect outdoor runs of coaxial cable, what type of surge suppressors

are commonly used?

14. The coaxial communications circuits follow the same Code requirements as which of the following:
    A. Electrical wiring B. Communication circuits
    C. Fiber optic cables D. A hangman's rope
15. Matching: Match each cable type with its use.
    A. CATVR (1) Any location except risers and plenums
    B. CATVX (2) Spaces that have environmental air, plenums, or ducts
    C. CATVP (3) Very fire-resistant; runs in installation shafts or floors in buildings
    D. CATV (4) Limited to only dwellings and raceways

CHAPTER 7

# NETWORK CABLING

Unlike telephone and video cabling, which have used one basic type of cable all along, computer networks have used many different types of cables, and in fact several other means of communication. All of these methods have been tried as solutions to providing low-cost, high-performance cabling. They include:

Unshielded twisted pair cables (UTP)
Shielded twisted pair cables (STP)
Coaxial cables
Optical fiber cables
Radio signals
Infrared light, reflected off ceilings or focused line of sight
Electronic signals sent through power lines

While most network cabling has migrated to UTP cabling following the structured cabling guidelines in EIA/TIA 568, there are still applications that use other cabling schemes. Therefore, a review of the entire spectrum of cables is needed to understand network cabling.

## CABLING REQUIREMENTS

With power work, we are most concerned about the path the current will take, and less concerned about the quality of the power going from one point to another. In data wiring, however, we must consider two qualities of the transmission.

First, we must have a clear path from one machine to the next. Here we are concerned with the signal's strength; it must arrive at the far end of the line with enough strength to be useful.

Second, we are concerned with the quality of the signal. For instance, if we send a square-wave digital signal into one end of a cable, we want a good square-wave signal coming out of the far end. If this signal is distorted, it is unusable, even if it is still strong.

The requirements are determined by a number of performance parameters of the cable, including its attenuation, characteristic impedance, capacitance, and crosstalk. Since we will use these terms many times in describing cables, we will take time to define them now.

### Attenuation

Attenuation (Figure 7–1) is the loss of signal power over the length of a cable. This attenuation is normally measured as a number of decibels per 100 feet, at a given frequency. For fiber optic work, attenuation is measured in decibels per kilometer.

Signals sent over copper wires deteriorate differently at different frequencies—the higher the frequency, the greater the attenuation. These losses come primarily from the capacitance of the cable, or from heat. Attenuation is a problem, since a weakened signal can only be picked up by a very sensitive receiver. Such very sensitive receivers are quite expensive; therefore, low attenuation is desirable.

Signal attenuation also depends on the construction of the cables, particularly the size of the wire and the dielectric characteristics of the cable insulation. For example, you could have two 100-ohm, 24-gauge cables and one of the cables might have a lower attenuation than the other, strictly because of the construction characteristics of the cables. This would allow the cable with lower attenuation to be used over longer distances with better results.

### Impedance

Impedance, which is the total opposition to current flow, is an important consideration for coaxial and twisted pair cables for computer systems. The most important consideration for impedance is that it remain consistent throughout the entire sys-

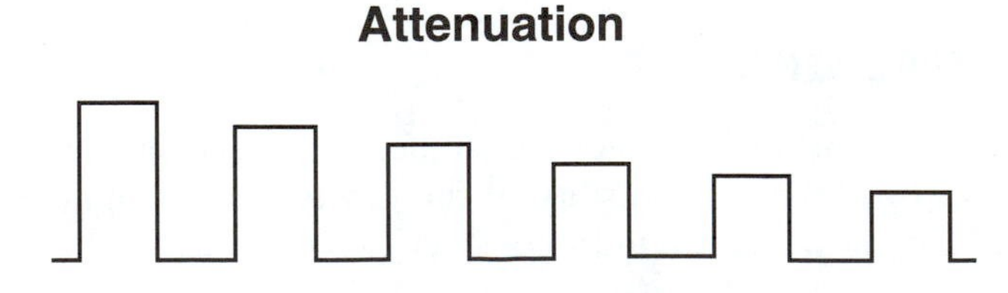


**Figure 7–1.** Signal attentuation is a decrease in amplitude.

tem. If it does not remain consistent, a portion of the data signals can be reflected back down the cable, often causing errors in the data transmissions.

### Cable Capacitance

As mentioned, capacitance is the most common element of attenuation. This capacitance usually comes in the form of *mutual capacitance*, which is capacitance between the conductors within the cable. Capacitance is a big factor because it causes the cable to filter out high-frequency signals. Data transmissions are square-wave signals. When the capacitance is too great, it has a tendency to round-off the digital data signals, making it more difficult to receive.

When data signals have a clear square wave, they are easily intelligible to the receiver. If, however, the signals become rounded, they can be confusing to the receiver, causing a high *bit error rate (BER)* or causing data to be received in error.

### Crosstalk

Crosstalk (Figure 7–2) is the amount of signal that is picked up by a quiet conductor (a conductor with no signal being transmitted over it at the moment) from other conductors that are conducting data. This signal is picked up through electromagnetic induction, the same principle by which transformers operate. Crosstalk contaminates adjacent lines and can cause interference, overloaded circuits, and other similar problems.

### Preventing Crosstalk

The traditional method to prevent crosstalk and electromagnetic interference is to place a shield around the conductors. The three principal types of shields are these:

1. Longitudinally applied metallic tape
2. Braided conductors, such as are commonly used in coaxial cables
3. Foil laminated to plastic sheets

**Crosstalk**

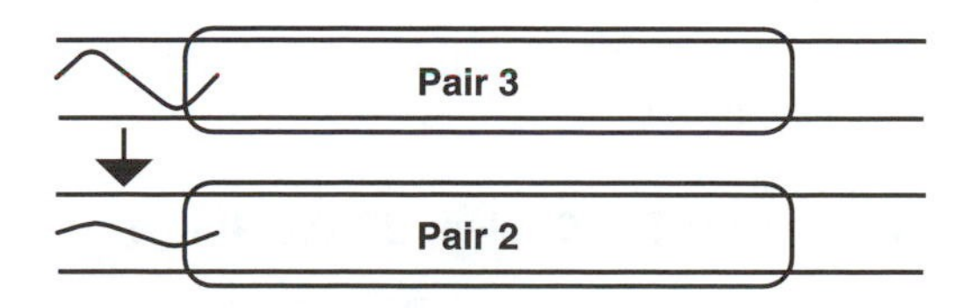


**Figure 7–2.** Crosstalk is signal coupling from one pair to another.

Spirally applied metallic tapes can also be used, but they are not generally good for data transmission.

Shielding stops interference and crosstalk by absorbing magnetic fields. Since the shielding is conductive itself, when the magnetic field crosses through it, it is absorbed into the shield. It does induce a current into the shield, but since this current is spread over the wide, flat surface of the shield, the field is diffused and is usually not strong enough at any one point to cause a problem.

Foil shields usually reduce electromagnetic interference by 35 decibels; wire braid shields generally reduce this type of interference by 55 decibels; and a combination of the two types of shields reduces interference by over 100 decibels. It is also possible to use electronics to reduce interference, although they must be closely matched to the network's data transmission rate.

Most of today's cables use twisted pairs and balanced transmission to minimize crosstalk. If the pairs are twisted at different rates, their antenna characteristics are different enough to greatly reduce crosstalk without any additional shielding.

## NETWORK CABLING TYPES

### Unshielded Twisted pair (UTP) Cables, 22–24 Gauge

*Applications:* Most current networks. UTP cable is the primary cable used for networks, as specified in the EIA/TIA 568 standard.
*Advantages:* Inexpensive; may be in place in some places; familiar and simple to install
*Disadvantages:* Subject to interference, both internal and external; limited bandwidth, which translates into slower transmissions unless multiple pairs or encoding electronics are used.

### Screened Twisted pair (ScTP) Cables

*Applications:* Same as UTP. While not currently specified for any networks or covered in the EIA/TIA 568 standard, it is used in many networks in Europe where EMI is a greater concern.
*Advantages:* A foil screen enclosing all four pairs helps electromagnetic intereference (EMI) emissions and interference, but has no effect on crosstalk
*Disadvantages:* More expensive, harder to terminate, require special plugs and jacks

### Shielded Twisted pair (STP) Cables, 22–24 Gauge

*Applications:* IBM Token Ring. STP is covered in EIA/TIA 568 but is no longer widely used.

*Advantages:* Easy installation; reasonable cost; resistance to interference; better electrical characteristics than unshielded cables; better data security
*Disadvantages:* Not easily terminated with modular connectors; may become obsolete due to technical advances

### Coaxial Cables

*Applications:* Original Ethernet
*Advantages:* Familiar and fairly easy to install; better electrical characteristics (lower attenuation and greater bandwidth) than shielded or unshielded cables; highly resistant to interference; generally good data security
*Disadvantages:* More expensive; bulky; may become obsolete due to technological advances

### Optical Fiber Cables

*Applications:* FDDI, ESCON, Fibre Channel, optional for most networks
*Advantages:* Top performance; excellent bandwidth (high in the gigabit range, and theoretically higher); very long life span; excellent security; allow for very high rates of data transmission; cause no interference and are not subject to electromagnetic interference; smaller and lighter than other cable types
*Disadvantages:* Slightly higher installed cost than twisted pair cables, since more expensive electronics interface to them.

## WIRELESS TRANSMISSION

In the past several years, there has been a good deal of interest in wireless networks. While part of the reason for this interest is due to the expense of installing cables (the installation expenses are usually far higher than the material expenses), and the frequent moving of terminals within a large office, the most compelling reason is the need to have mobile terminals connected to a network.

With a wireless network, no data transmission cables are required to connect any individual terminal. Within the range of the radio signals, a terminal can be moved anywhere. Wireless networks do require cabling, however, to connect the antennas to the network backbone, so they are not completely without cabling requirements.

Wireless networks are usually more expensive than cabled networks. Where they really save money is where terminals must be mobile, such as delivery services, warehouses, or hospitals. Money can also be saved in locations where it would be especially difficult to install cables.

The number one problem with wireless networks is their speed. It is very difficult to transmit data over available radio frequencies as fast as can be done over copper wires. To compensate for this impediment, some wireless networks have gone to multiple frequencies. By sending signals over several radio frequencies at the same time, signals can be sent far faster than over a single frequency. Special software is required to send these signals in a coded format.

A potential hazard with wireless networks is the health effects of radio emissions. Although no definitive studies have been completed, there have been indications that radio waves in these frequency ranges may be harmful.

## INFRARED TRANSMISSION

Another method of transmitting data signals without the use of wires is by using infrared (IR) light. By sending pulses of infrared light in the same patterns as electronic pulses sent over cables, it is possible to send data from one place to another. Networks based on IR transmission have been developed for use in offices and for line-of-sight transmissions between buildings.

Infrared light is used for this purpose because it is invisible to the naked eye, and because it is inexpensive to implement. This is a variation on the same technology we use for TV remote controls. The distance between terminals is normally limited to around 80 feet, although newer systems exceed that figure. Line-of-sight systems between buildings can go thousands of feet but are vulnerable to heavy rainstorms, fog, and other natural phenomena (including big birds)

This technology works fairly well, although there are problems that develop in offices with numerous walls. Just like normal light, infrared light cannot pass through walls. In open offices, this is not much of a problem, but in walled-off offices, remote transmitter/receivers are often required. A transmitter/receiver can be mounted in an area where it will easily receive data signals, and be connected to the computer with a cable. Not all of these systems use combination transmitter/receivers, however; some of them use separate transmitters and receivers.

As with other wireless signal transmission, IR networks require all of the parts that conventional networks require. Where the data cables would normally connect to the back of the computer, however, transmitter/receivers are installed instead. Infrared signals can be used to send and receive through these devices, usually at a rate of between 4 and 16 million bits per second. Depending on the design, the units can be rather large and awkward.

One great advantage of the wireless networks is that they can be set up anywhere, almost instantly. This can be a terrific feature for people whose work location is constantly shifting from one place to another. Some people in the trade have come to call these systems "networks in a box."

## POWER LINE CARRIER NETWORKS

Another method of sending data signals involves sending them through regular power lines. This method requires a special network adapter that modulates the computer data and transmits it over the power lines in an office or home. Other devices or computers connected the same way may receive data over the power lines. This system is very attractive from the standpoint of installation costs. No cables are required, and the installation is very quick and easy.

This system imposes data signals right over the power line current. The voltage and frequency differences are easily separated from each other by tuned receivers. (Power lines operate at 60 cycles per second and 120 volts, while data signals are normally in the 5-volt range and have a frequency of hundreds of thousands or millions of cycles per second.)

Power line carrier systems will never meet the technical performance of optical fiber cables, or even other types of cables. They are, however, a very appealing option for home networks, for example, where computers in the future may control heating and air-conditioning systems, turn lights on and off, and manage security systems.

The installation of these systems is not completely hassle-free, however. Power lines do not travel unbroken throughout whole buildings, or even through parts of buildings. To bridge the gaps in wiring systems, special devices called "signal bridges" are required. These devices connect to two separate wiring systems and transfer the data signals from one to another, without allowing current to be transferred from one system to another (which would cause major problems and, most likely, injuries). Signal amplifiers are also frequently required when wiring systems cover long distances. There can also be filters and other devices required for these systems, depending on the location of the installation.

## OTHER TRANSMISSION MEANS

While the methods of transmission that we have discussed are by far the most common methods, there are other methods that are sometimes used. The chief among these are microwave and laser signal transmission for line-of-sight connections between nearby buildings.

The typical method is to set up transmitter/receivers on the roofs of both buildings, and to connect both ends to the networks in the buildings. The buildings can be up to about one kilometer apart for most systems, and even further apart for others. Signal transmission rates of 1.5 MB/s (megabits per second) or greater are not uncommon.

It is important to remember in planning such a system, that the management of any buildings that are under the transmission area should be notified, and any appropriate permissions granted before any work is begun. Most microwave systems require FCC licenses, which may be difficult to obtain in metropolitan areas.

## *NEC*® CODE REQUIREMENTS

Article 725 covers Class 1, 2, 3, or 4, remote-control, signaling, and power-limited circuits. This article can be very confusing if you do not understand that circuits designated as Class 1, 2, 3, or 4 can be either *power-limited* circuits or *signaling* circuits. *Every circuit* covered by this article is Class 1, 2, 3, or 4. *Some* of them are power-limited, and *some* are signaling circuits.

It is important to note that this article does not apply to such circuits that are part of a device or appliance. It applies only to separately installed circuits.

### Definitions

One of the more difficult parts of this article is that it involves a lot of terms with which most of us are not familiar. The key terms are as follows:

**Class 1:** Circuits supplied by a source that has an output of no more than 30 volts (AC or DC) and 1,000 volt-amperes. (*Volt-amperes* [VA] is essentially the same thing as watts. You will find the term volt-amps used when relating to transformers and similar devices, since it is technically more correct for such uses than watts.) See section 725-11.

**Class 2:** Circuits that are inherently limited in capacity. They may require no overcurrent protection, or may have their capacity regulated by a combination of overcurrent protection and their power source. There are a number of voltage, current, and other characteristics that define Class 2 circuits. These characteristics are detailed in Tables 725-31(a) and 725-31(b).

**Class 3:** Class 3 circuits are very similar to Class 2 circuits. They are inherently limited in capacity. They may require no overcurrent protection, or may have their capacity regulated by a combination of overcurrent protection and their power source. The voltage, current, and other characteristics that define Class 3 circuits and differentiate them from Class 2 circuits are detailed in Tables 725-31(a) and 725-31(b).

**Class 4:** Class 4 circuits are Class 1 circuits that are also isolated, and they are used only for process control instrumentation.

**Power-limited:** This refers to a circuit that has a self-limited power level. That is, whether because of impedance, overcurrent protection, or other power source limitations, these circuits can only operate up to a limited level of power.

**Remote-control and signaling circuits:** These are Class 1, 2, 3, or 4 circuits that do not have a limited power rating.

**Supply side:** This is essentially the same thing as "line side." This term is used instead of "line" because of possible confusion. (Someone might think it referred to the power circuits, rather than simply the power that feeds the circuit.)

**CL2, CL3P, CL2R, and so on:** These are specific types of power-limited cables. See Table 725-50.

**PLTC:** Power-limited tray cable.

### Requirements

A remote-control, signaling, or power-limited circuit is the part of the wiring system between the load side of the overcurrent device or limited-power supply and any equipment connected to the circuit. There are many different requirements listed in article 725. The ones we will cover here will be the most important of the installation requirements.

We will begin with Class 1 requirements. Remember that these Class 1 requirements apply equally to Class 4 circuits; the only difference between Class 1 and Class 4 circuits is where they are installed. The requirements are as follows.

Except for transformers, the power supplies that feed Class 1 circuits must be protected by an overcurrent device that is rated 167 percent of the power supply's rated current. This overcurrent device can be built into the power supply; but if so, it cannot be interchangeable with devices of a higher rating. In other words, you cannot use any interchangeable fuse (as most are) in this power supply. Transformers feeding these circuits have no requirements except those stated in article 450.

All remote-control circuits that could cause a fire hazard must be classified as Class 1, even if their other characteristics would classify them as Class 2 or Class 3. This applies also to safety control equipment, but does not include heating and ventilating equipment.

Class 4 circuits cannot occupy the same raceway or cable with Class 1 circuits or power conductors. Class 1 circuits are not allowed in the same cable with communications circuits.

Power sources for these circuits must have a maximum output (note that this is *maximum output*, not *rated power*; the two terms are very different) of no more than 2,500 VA. This does not apply to transformers.

Class 1 remote-control and signaling circuits can operate at up to 600 volts, and their power sources need not be limited.

All conductors in Class 1 circuits that are #14 AWG or larger must have overcurrent protection. Derating factors cannot be allowed. Conductors that are #18 must be protected at 7 amps or less, and #16 conductors at 10 amps or less. There are three exceptions to this (see section 725-12).

Any required overcurrent devices must be located at the point of supply. See section 725-13 for two exceptions to this requirement.

Class 1 circuits of different sources can share the same raceway or cable, provided that they all have insulation rated as high as the highest voltage present. If only class 1 conductors are in a raceway, the allowable number of conductors can be no more than easy installation and heat dissipation will allow.

### Definitions of Class 2 and Class 3 Circuits

These are circuits that are inherently limited in capacity. They may require no overcurrent protection, or may have their capacity regulated by a combination of overcurrent protection and their power source. There are a number of voltage, current, and other characteristics that define these circuits. These characteristics are detailed in Tables 725-31(a) and 725-31(b).

If you refer to Tables 725-31(a) and 725-31(b) in the Code, you will find that there are quite a few combinations of circuit characteristics that can make the circuit Class 2 or Class 3. Note that the characteristics for AC and DC circuits are different. These tables have two particular groupings—circuits that require overcurrent protection, and circuits that do not. The only substantial difference between Class 2 and Class 3 circuits is that Class 3 circuits generally have higher voltage and power ratings than Class 2 circuits.

### Requirements of Class 2 and Class 3 Circuits

The basic requirements for Class 2 and Class 3 circuits are as follows.

Power supplies for Class 2 or 3 circuits may not be connected in parallel unless specifically designed for such use. The power supplies (when necessary) that feed Class 2 or 3 circuits must be protected by an overcurrent device that cannot be interchangeable with devices of a higher rating. You cannot use an interchangeable fuse in this power supply. The overcurrent device can be built in to the power supply. All overcurrent devices must be installed at the point of supply.

Transformers that are supplied by power circuits may not be rated more than 20 amps. They may, however, have #18 AWG leads, as long as the leads are no more than 12 inches long.

Class 2 or 3 conductors must be separated at least 2 inches from all other conductors, except in the following situations:

1. If the other conductors are enclosed in raceway, metal-sheathed cables, metal-clad cables, NM cables, or UF cables
2. If the conductors are separated by a fixed insulator, such as a porcelain or plastic tube

Class 2 or 3 conductors may not be installed in the same raceway, cable, enclosure, or cable tray with other conductors, except:

1. If they are separated by a barrier
2. In enclosures, if Class 1 conductors enter only to connect equipment that is also connected to the Class 2 and/or 3 circuits

3. In manholes, if the power or Class 1 conductors are in UF or metal-enclosed cable
4. In manholes, if the conductors are separated by a fixed insulator, in addition to the insulation of the conductors
5. In manholes, if the conductors are mounted firmly on racks
6. If the conductors are part of a hybrid cable for a closed-loop system. (See article 780.)

When installed in hoistways, Class 2 or 3 conductors must be enclosed in rigid metal conduit, rigid nonmetallic conduit, IMC, or EMT. They can be installed in elevators as allowed by section 620-21. In shafts, Class 2 or 3 conductors must be kept at least 2 inches away from other conductors.

Any cables that are used for Class 2 or 3 systems must be marked as resistant to the spread of flame. All Class 2 and 3 cables also are marked with a listing of the areas in which they may be installed. The cables may be installed only in the listed areas.

Two or more Class 2 circuits can be installed in the same enclosure, cable, or raceway as long as they are all insulated for the highest voltage present. Two or more Class 3 circuits may share the same raceway, cable, or enclosure. Class 2 and Class 3 circuits can be installed in the same raceway or enclosure with other circuits, provided the other circuits are in a cable of one of the following types:

Power-limited signaling cables (see article 760)
Optical fiber cables (see article 770)
Communication cables (see article 800)
Community antenna cables (see article 820)

When Class 2 or 3 conductors extend out of a building and are subject to accidental contact with systems operating at over 300 volts to ground (not 300 volts between conductors, but 300 volts to ground), they must meet all the requirements of section 800-30 for communication circuits.

## NETWORK CABLE HANDLING

The performance of the cabling network is heavily dependent on the installation. The Category 5 components used in most structured cabling installatins have been carefully designed and exhaustively tested to meet or exceed the requirements of EIA/TIA 568 and TSB-67 for performance at 100 MHz. If the cable is not properly installed, performance will be degraded. You cannot mix Cat 5 and Cat 3 components either; every component in the cable plant must be Cat 5 rated and properly installed or the cabling will fail testing.

As general guidelines, remember the following:

- All components must be Cat 5 rated for Cat 5 performance.
- Cable must be pulled from the reel or box without kinking.
- Cable must be pulled with less than 25 pounds of tension.

- Use cable lubricant in conduit if necessary.
- Cable must not be pulled around sharp corners or kinked.
- Inspect the cable routes for surfaces that may abrade the cable.
- On riser installations, try to lower the cable down, not pull up.
- Cables must be supported to prevent stress. Cable supports should not have sharp edges that may distort the cable.
- Cable ties must not be so tight as to distort the jacket of the cable. They are only used to prevent unnecessary movement of the cable, so snug is tight enough.

## TYPICAL INSTALLATIONS

The typical installation of communications cabling is in an office building. There are communications wiring closets on every floor, and several if the floor covers a large area. A main communications closet will have the primary communications equipment (PBX, network routers, and so on). Cables will run from the wiring closet to work areas overhead above a suspended ceiling. Wiring closets will be connected by multiple cables run overhead or in risers.

Each work area (Figure 7–3) will be connected to the closet with at least two cables, one for telephone and one for data. While the telephone cable can be Cat 3, it is more common to install two Cat 5 cables for simplicity and easier cabling man-

**Figure 7–3.** Work area outlet for network and telephone in modular office.

agement. If the telephone is already installed and only a data cable is being installed, only one cable may be needed. Cables will be terminated in 8-pin modular jacks at the outlet and either in patch panels (Figure 7–4) or on punchdown blocks in the closet.

Closets may have numerous cables running between them, unless the backbone cabling is fiber optic, which will be only one cable with several fibers. Numerous cables will be needed to connect wiring closets to the equipment room. Terminations may be either punchdowns or patch panels or both, depending on the destination of the cables.

Some applications will have cables in conduit to protect cables in areas where damage is possible, to prevent electromagnetic interference (EMI) where that is a problem, or to assist in firestopping. Cable trays (Figure 7–5) may be used to keep all cables neatly together and out of harm's way. Some installations may have false floors to allow interconnection cables to be run below the floor.

**Figure 7–4.** Terminated Cat 5 cables in modular patch panels.

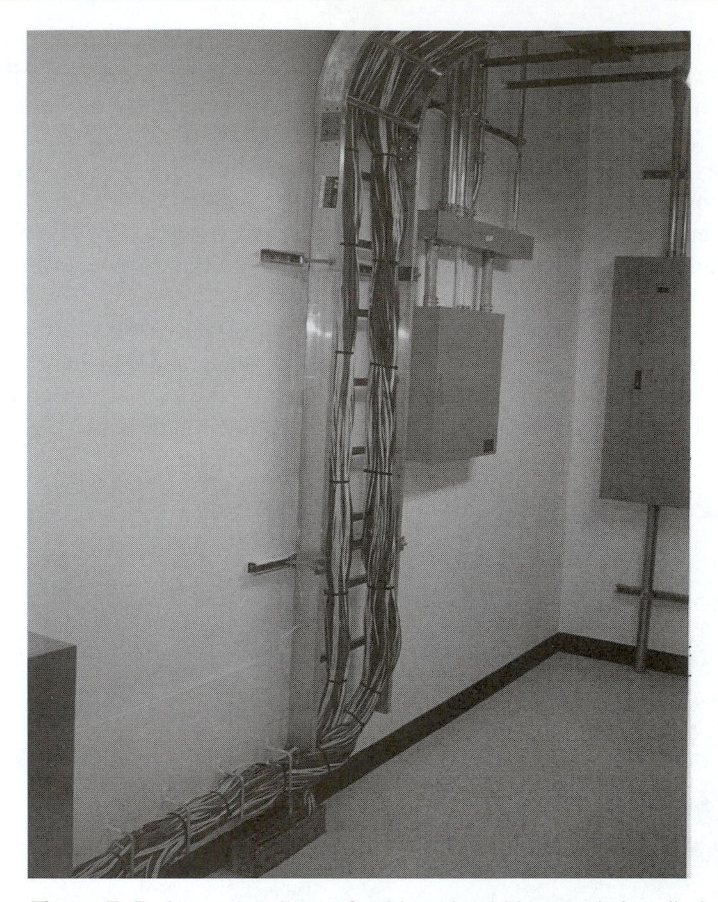

**Figure 7–5.** Large numbers of cables should be neatly bundled and placed in cable trays.

## PULLING CABLES

### Horizontal Pulls

In a typical installation, the horizontal cables will be pulled between the wiring closet and the work area above a suspended ceiling. This pull will involve routing the cables around everything else already there, including other cables, HVAC (heating, ventilating, and air-conditioning) systems, light fixtures, and so on. It is usually easier in new construction, since the tiles are not in place. The procedure involves gaining access, determining the best route for the cables, running a handline, affixing cables to the handline, and pulling the cables.

The most efficient way to pull horizontal cables is to pull bundles. Using drawings of the area, determine where all work areas are located and establish several consolidation points or clusters from which to pull. This is easy with open areas and modular furniture, but may be much more difficult in walled offices. Sometimes the best way to pull cables is from the work area (sometimes called the "drop") to the closet, but sometimes pulling from the closet is more efficient. It all depends on the actual installation situation.

The first step is to gain access to the area. Clear as many obstacles as possible and secure the area. Put caution tape around the ladder you are working on, and keep everyone away for safety (theirs as well as yours). In the path you will be taking with the pull, carefully remove every other ceiling tile along the route, but leave them above the ceiling to prevent damage.

After examining the area and finding the most direct route, tie a weight to the handline or pull string (typically a thin nylon rope) and, starting at the drop end, toss it from ceiling opening to opening until you reach the closet. Some installers even use a slingshot to shoot the handline over ceiling tiles.

Cables come in several styles of boxes designed to allow easy pulling directly from the box (Figure 7–6). Note the instructions on the box regarding the proper placement of the box for pulling to ensure the cable is not twisted or kinked when pulled from it. Place all the cable boxes in the drop area, and mark each as to the final location. Post-it® notes are good temporary labels for the boxes.

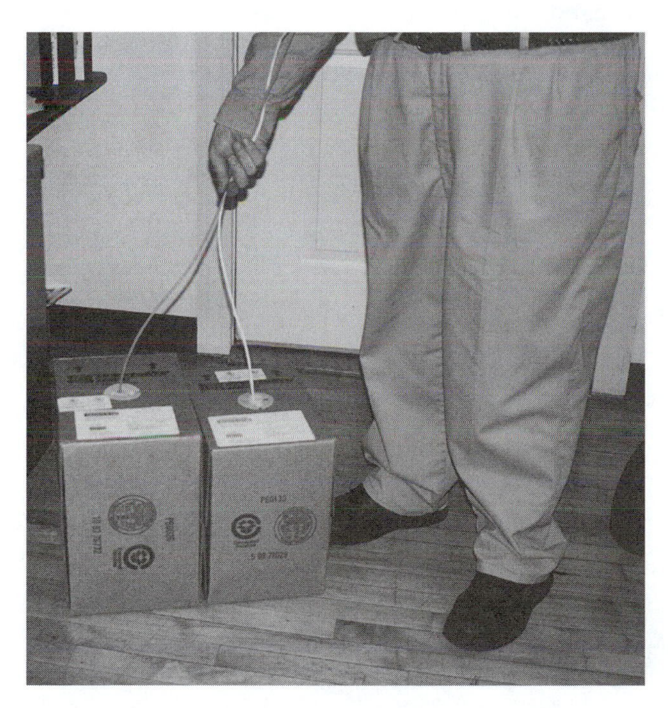

**Figure 7–6.** Boxed UTP cable is made to be pulled directly from the box.

If the cables are on reels, set up a stand with an axle to allow the cable to unroll. Do not feed the cable off the sides of a spool, as it will twist and kink. Two chairs and a short length of conduit or a broomstick can be a makeshift axle, and some contractors use jack stands and a piece of conduit. Cable "trees" are also available at reasonable prices.

Mark the cables with a permanent marker with the same identification at both ends (Figure 7–7). This will make later identification of the cable easier and make tracing unnecessary.

Although cables can be pulled individually, it is more efficient to pull as many cables as possible at one time. Bundle the cables (Figure 7–8) as follows:

- Gather all the cables to be pulled at once and align the ends. Tape them together with electricians tape for 8 to 10 inches, including the ends.
- Split the bundle in half.
- Run the handline between the two bundles and tie it to the handline. Tape the handline to the tip. This makes the bundle less resistant to pulling.
- Some installers like to make loops on the end by taping the bundle about a foot back from the end, dividing the cables, tying them in a knot to form a loop on the end, and taping it. The handline can be attached to that loop.

If you are pulling only one cable, make a loop in the end, twist the cable back over itself, tape it, and tie the handline to the loop (Figure 7–9).

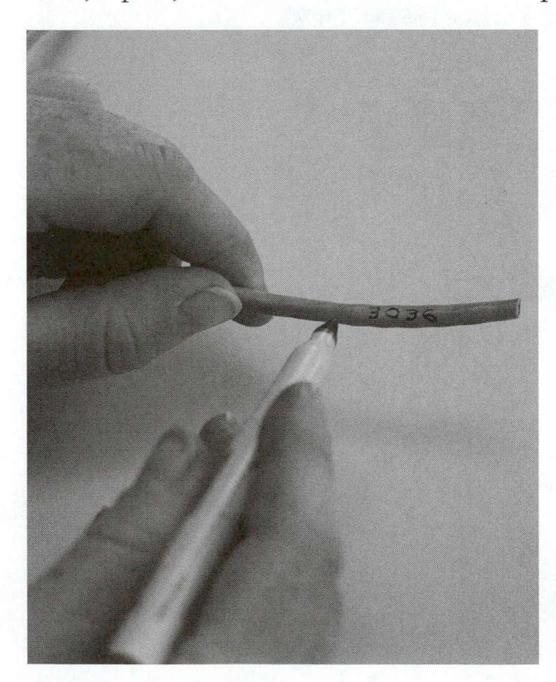

**Figure 7–7.** Mark the cables with a permanent marker to facilitate identification.

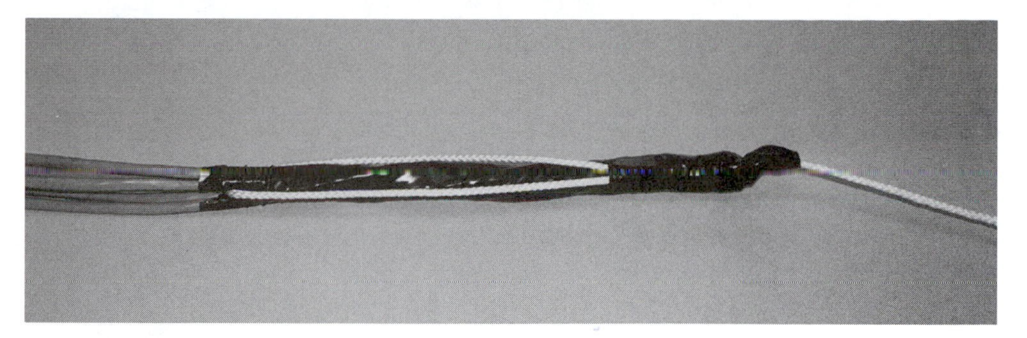

**Figure 7–8.** Bundles of cable ready for pulling.

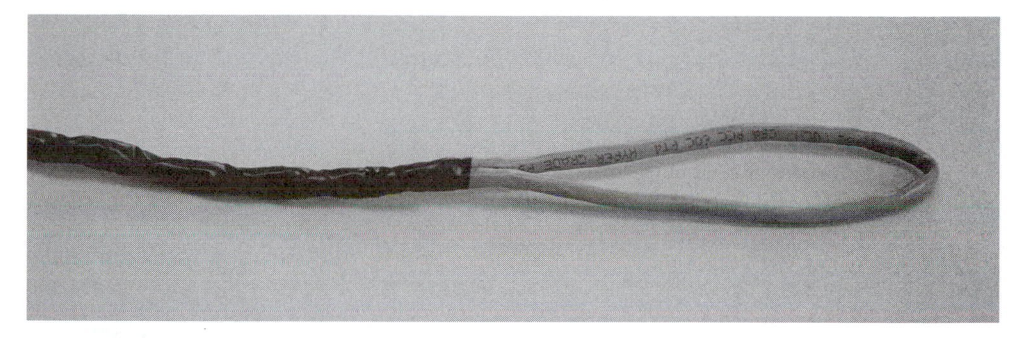

**Figure 7–9.** A single cable with a loop for pulling.

Pull the cable, watching for snags, as you pull it around sharp corners or other obstacles. If you hit a snag, stop and free the cable. Never jerk the cable, as that may overstress it and cause performance degradation. Find the cause of the snag and fix it!

### Horizontal Cable Supports

After pulling the cable, it must be supported. It is not safe to leave it laying on the ceiling; it must be supported and kept away from the AC power cables and lighting fixtures. Numerous hangers, cable trays, and other cable handling solutions are available from vendors of cabling hardware.

Above suspended ceilings, "J" hooks can be used to separate the cables from most other hardware above the ceiling, especially power cables and lighting fixtures. Special J hooks with wide bases should be used for communications cables (Figure 7–10). Hangers can sometimes be attached to the ceiling where the suspended ceiling hangers are attached, but J hooks should not be attached to the ceiling suspension wires. Using suspended ceiling tile wires is not allowed in most localities; such

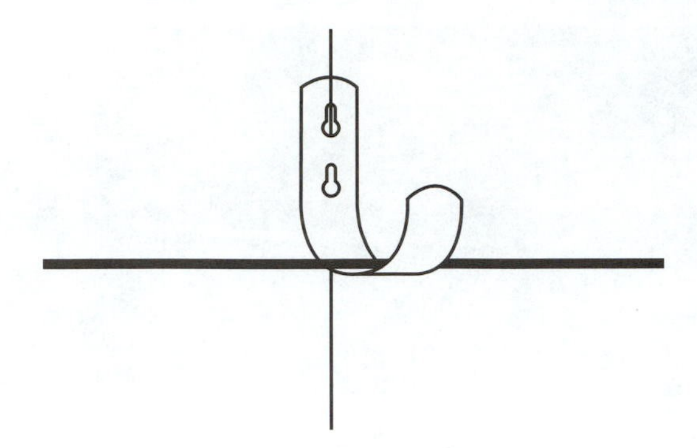

**Figure 7–10.** Special wide "J hooks" should be used for Cat 5 cables.

wires may cause the ceiling to become uneven. In new or renovated construction, it may be possible to have separate wires installed for cable J hooks at the same time as the ceiling support wires are installed.

### Closet-to-Closet Pulls

Pulls from closet to closet or equipment rooms can be simple pulls of cable bundles like horizontal pulls or may involve conduit, cable trays, or riser installation. The first consideration is the same as that in all installations—examine the route closely, and plan the pull accordingly.

### Conduit Pulls

Make sure you know how large and long the conduit is and how many bends are involved. The diameter of the conduit will limit the number of cables that can be installed, and the current length and bends of the cables will dictate whether lubrication is necessary. For long runs, use special cable lubricants only, as other lubricants may damage the cables.

Conduit pulls follow this procedure:

1. Open the conduit on each end. Examine the ends for roughness, and use a special "leaderguard" (Figure 7–11) if necessary to prevent cable damage.
2. Feed a snake into the conduit from the end you will be pulling the cables into.
3. At the far end, attach a handline to the snake and pull back through the conduit.
4. Set out all cable boxes and mark each by number or location. Arrange so the cables will pull smoothly without kinking or twisting.
5. Bundle the cables as described in the section on horizontal pulls.
6. Attach the handline to the cable bundle.

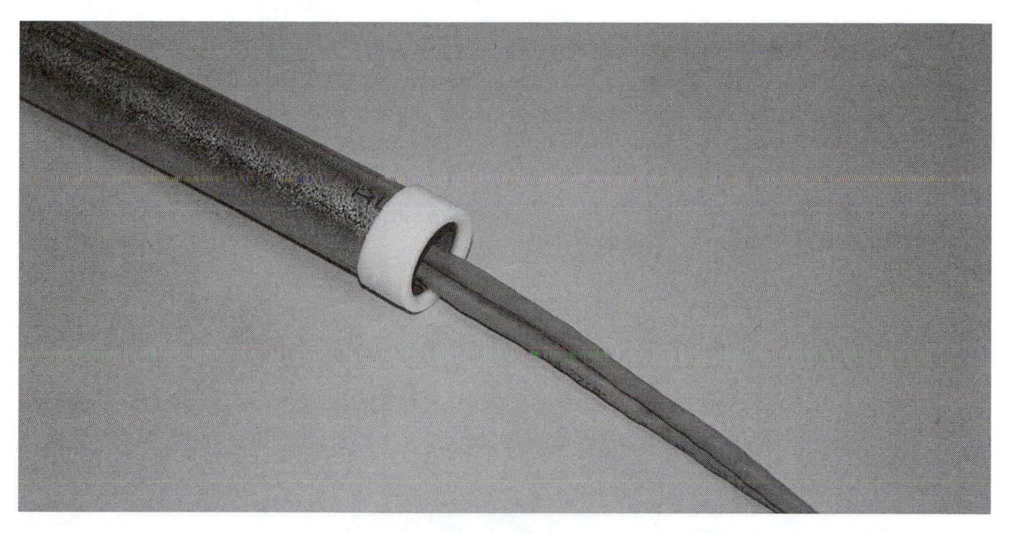

**Figure 7–11.** Plastic guards protect cable in conduit.

7. With one operator at the beginning feeding cables and one or more pulling on the far end, pull the cables slowly and smoothly through the conduit.
8. Use mechanical pullers if the tension is too high. They can pull more smoothly and at higher tensions, but should have tension monitors to prevent overstress.

### Cable Trays and Raceways

Cable trays (Figure 7–12) are used to make installations neater and to protect cables from sharp edges and abrasive surfaces and from having other cables or hardware put stress on them. In order to protect the cables, it is necessary to arrange the cables properly in the trays. Lay the cables flat in the trays, starting from the sides and avoiding sharp bends at corners. If fiber optic and copper cables are placed in the same trays, the fiber cables should be on top.

### Riser Installation

Often closets or equipment rooms will be on different floors. For short runs, the cable can be supported from the top with a mesh grip or similar hardware available from cabling hardware vendors. For longer runs, special cables may be used or a "messenger," a steel cable to which the cable can be attached, is installed with anchors at top and bottom (and along the way if needed) to which the cables are attached.

**Figure 7–12.** Overhead cable trays organize and protect cables.

Although we call it a riser "pull," it is much easier to drop the cable down from the top and let gravity help you. Like all other pulls, make sure you know the exact path of the cables and ensure that the path is clear. Make sure you know where and how the cable will be attached. You do not want to lose the cables!

If needed, use a cable pulley above the riser opening. This is sometimes necessary if the cable is being dropped through a large opening and needs guidance.

Riser installation procedures are as follows:

1. Set up the cable or cable bundle about 30 feet from the riser opening.
2. Have as many people as needed holding the cables and helping feed them into the riser opening.
3. Have personnel at each location where the cable needs guiding, such as each floor in a multifloor drop or where the cable will be attached to a messenger.
4. Slowly feed the cable into the opening, guide it through each floor, and proceed until the drop is complete.

## MOUNTING HARDWARE IN CLOSETS OR EQUIPMENT ROOMS

The closet or equipment room can be anything from a broom closet with communications hardware mounted on the wall to giant rooms with raised floors full of racks of equipment and cable trays connecting them. This makes describing this part of the installation difficult, so we will offer some generalities.

The layout of the room will be dictated by the space available and the equipment to be installed. Someone will lay out the area and specify where all the hardware will be installed. The installation of cable trays, racks, and so on is so vendor-specific that the best advice is read the directions.

All hardware should be installed before any cabling begins. Once all the hardware is installed, the cable can be pulled in and terminated at the proper locations.

### Firestopping

All penetrations of firewalls require firestopping to meet fire codes. This can be done with permanent foam-in firestopping material or removeable material available in bags if more cables are to be installed in the future (Figure 7-13).

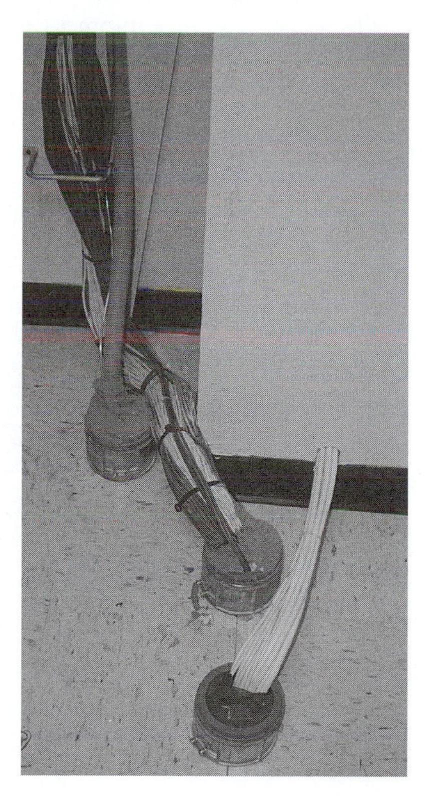

**Figure 7–13.** Penetrations of firewalls or floors must be firestopped to meet building codes.

## CABLING IN THE CLOSET OR EQUIPMENT ROOM

First and foremost, be neat. If this is a new installation, remember that others will be inspecting your work and later doing moves and changes. Run cables neatly and carefully. Avoid sharp bends that hurt cable performance. If you have excess cables, coil them up out of the way and tie them loosely. Mark everything and document where everything goes.

Cable ties are often used to hold bundles of cables neatly. Normal nylon cable ties are acceptable as long as they are not tied too tightly, which can cause problems with crosstalk or attenuation in Cat 5 cables or loss in fiber optic cables. If they are used, they should be hand-tightened only until they are snug, but can still be easily slid along the cable, then cut off. A better choice, although more expensive, are ties made from hook and loop fabric, which cannot harm the cables and offer the added advantage of being easily opened for adding additional cables (Figure 7–14).

If you are installing the cable hardware, you will usually be attaching all the punchdowns on plywood backboards attached to the wall. Again cables should be routed neatly, avoiding sharp bends or stresses, and held in guides or wire saddles. Follow the guidelines on terminating to preserve cable performance.

Patch panels are mostly rack-mounted. Cables must be brought to the racks, terminated, and dressed neatly. Patchcords will be used by the equipment installers to attach communications hardware to each location. Obviously, neatness is preferred, but it is sometimes difficult as more patchcords are added to the racks (Figure 7–15). The hook and loop type of cable tie can be used to hold coiled up excess patchcord cable for neatness. Good documentation is mandatory to speed the setup and moves and changes.

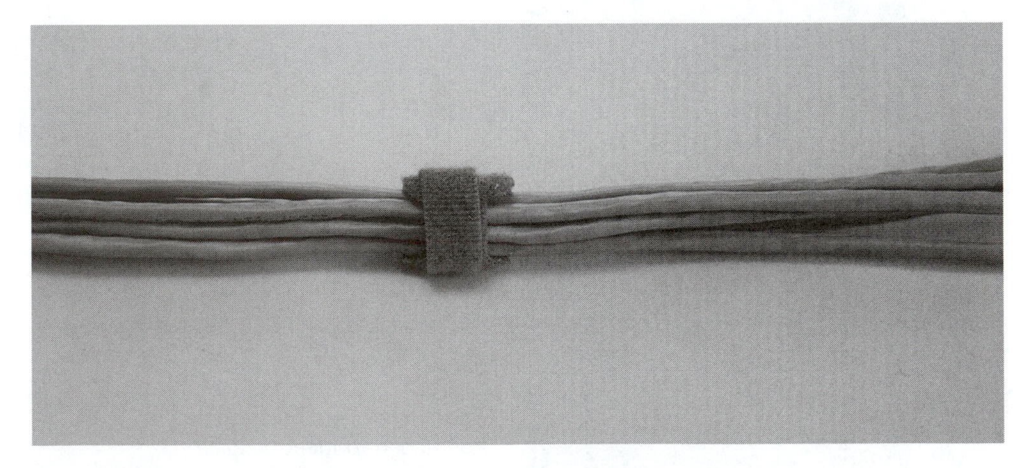

**Figure 7–14.** Cable ties made from hook and loop fasteners keep cables neat.

**Figure 7–15.** After many moves and changes, even the neatest patch panel will probably look like this!

## CHAPTER REVIEW

1. What two qualities does one concern oneself with in cable requirements of network cabling?
2. What is attenuation, and how is it specified in copper and fiber?
3. What affects attenuation in copper cables?
4. If the opposition of current flow, impedance, is not consistent, what happens to the signal?
5. What does capacitance filter out?
6. What does a distorted or rounded data signal cause?
7. Electromagnetic induction causes what type of problem?
8. List the three types of shields used to prevent crosstalk.
9. What type of transmission is used to minimize crosstalk in unshielded cables?
10. *Matching:* Match each wire with the appropriate application, advantage, and disadvantage. Each of the wires listed will have three numbers indicating these elements.

A. Coaxial cables
B. Unshielded twisted pair (UTP) cables
C. Optical fiber cables
D. Screened twisted pair cables
E. Shielded twisted pair cables

Applications:
(1) Most current networks
(2) Original Ethernet
(3) Same as UTP
(4) Optional for most networks
(5) IBM Token Ring

Advantages:
(6) Inexpensive
(7) Easy installation
(8) Better data security
(9) Excellent bandwidth
(10) Has no effect on crosstalk

Disadvantages:
(11) Hard to terminate
(12) Limited bandwidth
(13) Higher cost
(14) Bulky
(15) Breaching security

11. What type of cabling is needed for wireless transmission?
12. What is the advantage of wireless networks?
13. You can use Cat 3 and Cat 5 components together, since the wires are so similar in color and size. T or F
14. What type of terminations are used in wiring closets?
15. What is the procedure in pulling horizontal cables?
16. Once you have each box of cable in the drop area, what should you do to each box?
17. How do you support the wires above the ceiling?
18. When you run cable through a conduit, what can you use to protect the cable from the rough area at the end where the conduit was cut?
19. When placing both fiber and copper cables in a cable tray, which type of cable is placed on the bottom?
20. For long-run riser installations, you would use special cables or install a ______.
21. What will help in a riser "pull" to get the cable to the end destination?
22. What should be done in the closets or equipment room before pulling the cable?
23. What will help speed the process of a setup in a closet or equipment room?

CHAPTER

# 8

# TESTING VOICE, DATA, AND VIDEO WIRING

All wiring installed for voice, data, or video applications needs testing to ensure proper operation. Data and voice cabling are similar in testing requirements and are covered in one section. Coax testing can be either easier (if it works) or much harder (if it does not), so we will cover it first.

## COAX CABLE TESTING

Since coax has only two conductors, the inner conductor and the shield, you want continuity and no shorts. This can be tested by a simple coax tester or a digital multimeter (DMM). With the DMM, testing the cable with no terminations will tell if the cable is shorted, and testing with it terminated will tell whether or not it is open. If the cable passes these tests, it should work with most applications.

Sometimes coax can be damaged in installation and will have unusual attenuation characteristics over its frequency range. If a coax cable has problems transmitting a signal but shows neither an open nor a short, look for damage like kinks that may be causing the problem.

## UNSHIELDED TWISTED PAIR (UTP) TESTING

Testing voice and data UTP wiring is a function of use. Unless the network running on the wire is high speed, the wire only needs testing for correct connections, using a low-cost wiring verifier (wire mapper) or a "toner." As the network speed increases, the need for testing bandwidth, crosstalk, and so on at appropriate network speeds becomes more important.

Voice cabling requires a simple verification of the connections, or wire mapping. This is done with an instrument that plugs into the cables at either end and tests for proper pin-to-pin connections. A toner is a device that puts a tone on the wires and allows checking at any point with a small receiver. Toners can be used to trace and identify wires in termination blocks also.

Even though voice wiring may not require more than a continuity check, many structured wiring installations pull the same Cat 5 cabling for both voice and data applications, since the incremental cost of Cat 5 over Cat 3 is low and the installation of Cat 5 everywhere gives the most future flexibility. Therefore, it may be appropriate that all new wiring be fully tested as though any link may eventually have to carry network data.

Most testing today, with the exception of simple wire mapping, is done with automated loss testers that are preprogrammed to perform tests to the requirements of EIA/TIA TSB-67. These instruments test wire mapping, length, attenuation, and end crosstalk (NEXT). As long as the instruments meet the TSB-67 requirements and are designed for testing the networks proposed for the cable plant you are installing, you can expect valid test results, and can generally accept "pass/fail" results.

This discussion focuses on the tests used, what causes problems, and how to troubleshoot testing failures. If you are using an automated tester, read the directions and practice with the instrument before using it in the field, as that will make mistakes in interpreting data much less likely.

Table 8–1 shows the recommended testing requirements of different cables when used for different communications applications.

## WIRE MAPPING

Wire mapping includes all tests for correct connections. The correct pairs of wires must be connected to the correct pins, according to the color codes defined by the standards (either TIA T568A or T568B mapping). There must be no transposing of wires or pairs and no shorts or opens.

*Note:* As this book is being written, the EIA/TIA 568 standard committee is working on standards for an extended-performance Category 5 cabling system (Cat 5E). Additional specifications are propsed for testing for powersum crosstalk, return loss, far-end crosstalk, propogation delay and delay skew. For an update on the development of these specifications, talk to component manufacturers, tester manufacturers or refer to the Cable University website: www.cableu.net.

Figure 8–1 shows EIA/TIA T568A, and Table 8–2 gives a wire map for T568A, pin assignments. Figure 8–2 is another way to look at the wiring diagram for the 568A modular jack.

**Table 8–1** Network Cabling Test Requirements

| Cable/Network | Cat 1, 2 | Cat 3 | Cat 5 |
|---|---|---|---|
| Analog phone | W | W | W |
| Digital phone/PBX | Cat 1:NR, Cat 2:W | W | W |
| Ethernet | NR | L, W, X, A | L, W, X, A |
| 4-MB Token Ring | NR | L, W, X, A | L, W, X, A |
| 16-MB Token Ring | NR | L, W, X, A | L, W, X, A |
| 100Base-T4 | NR | NR | L, W, X, A, PD, S |
| 100Base-Tx | NR | NR | L, W, X, A, PD |
| 100VG AnyLAN | NR | L, W, X, A, S | L, W, X, A, S |
| TP-PMD/FDDI | NR | NR | L, W, X, A |
| 155 MB/s ATM | NR | NR | L, W, X, A |
| GB Ethernet | NR | NR | Undefined |

Tests: L = length, W = wire map of connections, X = NEXT (near-end crosstalk), A = attenuation, S = delay skew, PD = propagation delay, NR = not recommended

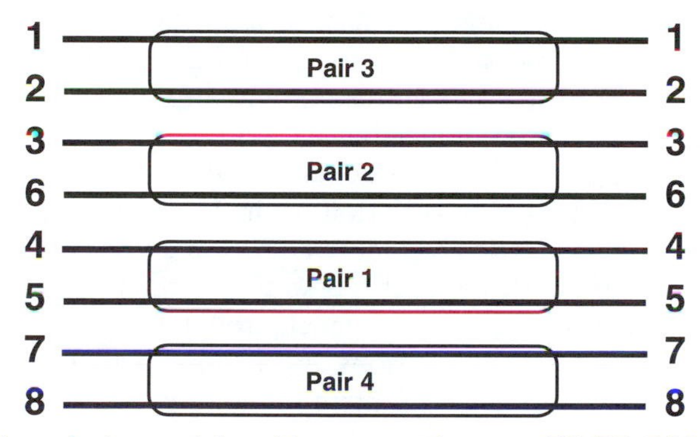


**Figure 8–1.** Eight-conductor modular wiring connections per EIA/TIA 568.

**Table 8–2** TIA T568A Modular Connector Wiring

| Pin Number | Pair Number | Color Codes |
|---|---|---|
| 1 | 3-Tip | White/Green |
| 2 | 3-Ring | Green |
| 3 | 2-Tip | White/Orange |
| 4 | 1-Ring | Blue |
| 5 | 1-Tip | White/Blue |
| 6 | 2-Ring | Orange |
| 7 | 4-Tip | White/Brown |
| 8 | 4-Ring | Brown |

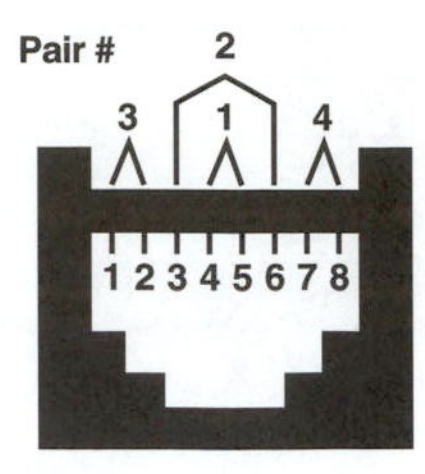


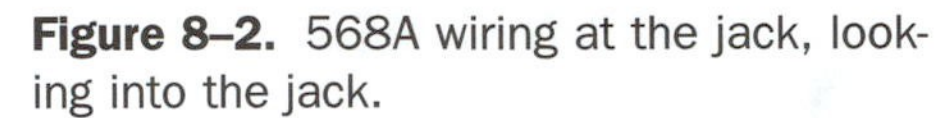
**Figure 8–2.** 568A wiring at the jack, looking into the jack.

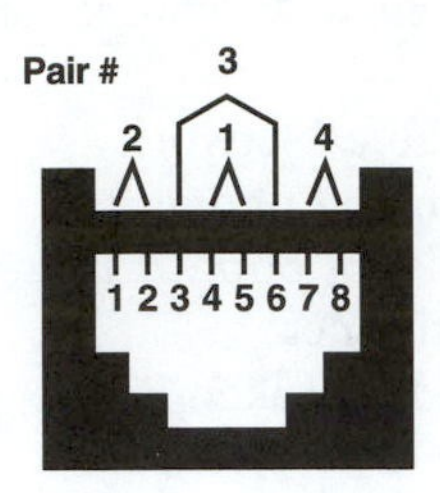


**Figure 8–3.** 568B wiring switches the position of pairs 2 and 3.

The 568A or 568B standard may be used for building wiring. The T568B standard (Figure 8–3) is more commonly specified by AT&T. The only difference between T568A and T568B wiring is the reversal of pairs 2 and 3, which makes no difference in performance—it is simply a different color code convention adopted by various vendors. (Table 8–3 gives a wire map for T568B pin assignments.) Whichever version is used, it must be used throughout an installation to prevent problems. Some plugs and jacks include both 568A and 568B wiring guides on the part, so be careful that one end is not wired to a different standard. If such a problem is detected, it can be solved by an equally miswired jumper, but this invites future problems. Better to correct it when detected!

There is yet another variation of the 8-pin pinout, the FCC USOC (Universal Service Order Code) RJ61X (Figure 8–4). Table 8–4 gives a wire map for USOC pin assignments. While this pinout works for voice and some low-speed data applications, the pair layout causes too much crosstalk for higher-speed networks, so it is seldom used for data installations and is not covered in the 568 standard. It does allow a modular 6-pin plug to attach to an 8-pin plug, although that is not recommended.

**Table 8–3** TIA 568B Modular Connector Wiring

| Pin Number | Pair Number | Color Codes |
|---|---|---|
| 1 | 2-Tip | White/Orange |
| 2 | 2-Ring | Orange |
| 3 | 3-Tip | White/Green |
| 4 | 1-Ring | Blue |
| 5 | 1-Tip | White/Blue |
| 6 | 3-Ring | Green |
| 7 | 4-Tip | White/Brown |
| 8 | 4-Ring | Brown |

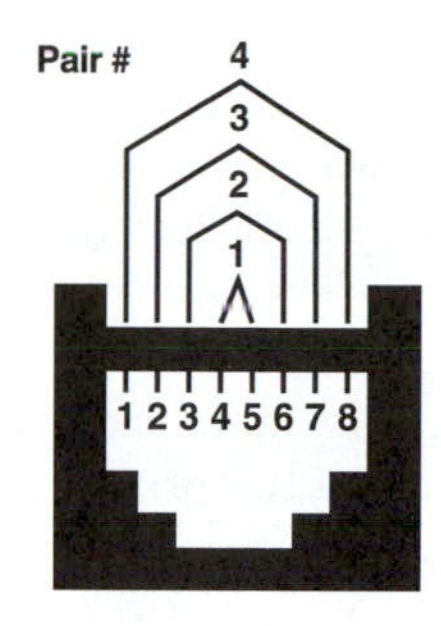


**Figure 8–4.** USOC wiring for the 8-pin modular jack is not used for LAN wiring, only.

**Table 8–4** USOC Modular Connector Wiring

| Pin Number | Pair Number | Color Codes |
|---|---|---|
| 1 | 4-Ring | Brown |
| 2 | 3-Tip | White/Green |
| 3 | 2-Tip | White/Orange |
| 4 | 1-Ring | Blue |
| 5 | 1-Tip | White/Blue |
| 6 | 2-Ring | Orange |
| 7 | 3-Ring | Green |
| 8 | 4-Tip | White/Brown |

Open DECconnect is a variation of T568A that leaves out pair 1. (Table 8–5 gives a wire map for DEC connect pin assignments.) It has only been used in Digital Equipment Corporation's proprietary DECconnect networks.

When EAI/TIA 568 was published, the T568A convention was recommended because its pin/pair assignments, with pairs 1 and 2 located on the center four pins,

**Table 8–5** Open DEC connect Wiring

| Pin Number | Pair Number | Color Codes |
|---|---|---|
| 1 | 3-Tip | White/Green |
| 2 | 3-Ring | Green |
| 3 | 2-Tip | White/Orange |
| 4 | NC | |
| 5 | NC | |
| 6 | 2-Ring | Orange |
| 7 | 4-Tip | White/Brown |
| 8 | 4-Ring | Brown |

were compatible with a wide variety of existing 2-pair voice and data applications including some data systems that used the old USOC pair configuration for 6- and 8-conductor jacks. T568B was an accepted alternative, for it was a convention that AT&T had already widely established for 4-pair data wiring. T568B is the same as AT&T or WECO 268A. Federal government publication FIPS 174 only recognizes T568A.

Most important for any cable is correct connections—that is, the proper conductors must be connected to the proper pins on the plugs and jacks. But first you must be certain whether you are dealing with T568A, T568B, USOC, or even DECconnect terminations, as opposed to nonstandard connections or older installations where less than four pairs are terminated, to get the correct test results.

Networks may not use all four pairs. In fact, most use only two pairs chosen to allow proper performance for that network. Table 8–6 gives a rundown of what pins are used by most common networks. You must translate that into "pairs," using Tables 8–2 through 8–5, depending on the pin configuration.

### Wire map Problems

Most wire map problems occur at the connections. Physical examination of the connections should find the fault. Wire map errors fall into several basic categories, which are illustrated with T568A connections in the following text.

*Shorts and opens.* A short is where two conductors are accidentally connected, and an open is where one or more wires are not connected to the pins on the plug or jack (Figure 8–5). Opens can also occur due to cable damage. A time domain reflectometer (TDR) test, which shows the distance to the fault, can assist you in locating the fault.

**Table 8–6** Network Conductor Use

| Network | Pins Used |
|---|---|
| 10Base-T | 1-2, 3-6 |
| Token Ring | 4-5, 3-6 |
| TP-PMD (FDDI) | 1-2, 7-8 |
| ATM | 1-2, 7-8 |
| 100Base-TX | 1-2, 3-6 |
| 100Base-T4 | 1-2, 3-6, 4-5, 7-8 |
| 100VG-AnyLAN | 1-2, 3-6, 4-5, 7-8 |
| Gigabit Ethernet or 1000Base-T | 1-2, 3-6, 4-5, 7-8 |

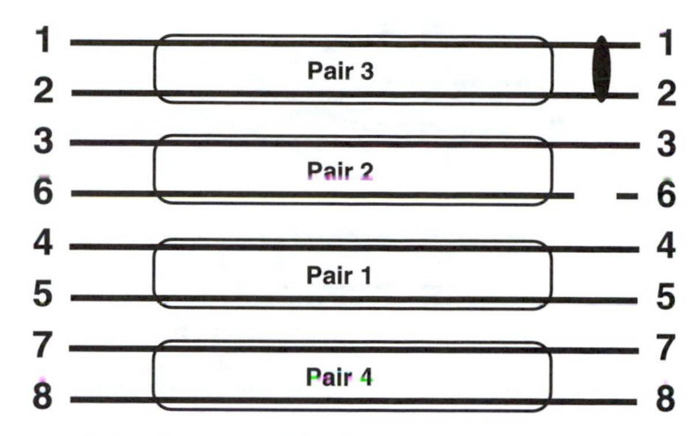


**Figure 8–5.** Opens and shorts at a termination.

*Reversed pairs.* Reversed pairs occur when the conductors (Tip and Ring) are reversed in the pair (Figure 8–6).

*Transposed or crossed pairs.* Transposed or crossed pairs occur when both conductors of one pair are reversed with both conductors of another pair at one end (Figure 8–7).

*Split pairs.* A wiring verifier will provide basic connection information, but some faults like split pairs (Figure 8–8) may not show up in a wire map. Split pairs occur when one wire on each of two pairs is reversed on *both* ends. This fault is impossible to find with a normal wiring verifier, since the wire map is correct—that is, the pin connections are correct, but the wires are not in proper pairs. Split pairs can be detected only in a crosstalk (NEXT) test, where the unbalanced pairs can be detect-

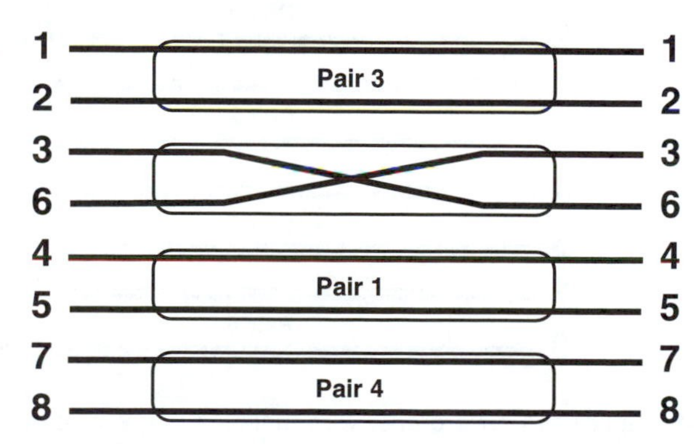


**Figure 8–6.** Reversed pair has conductors crossed at one connection.

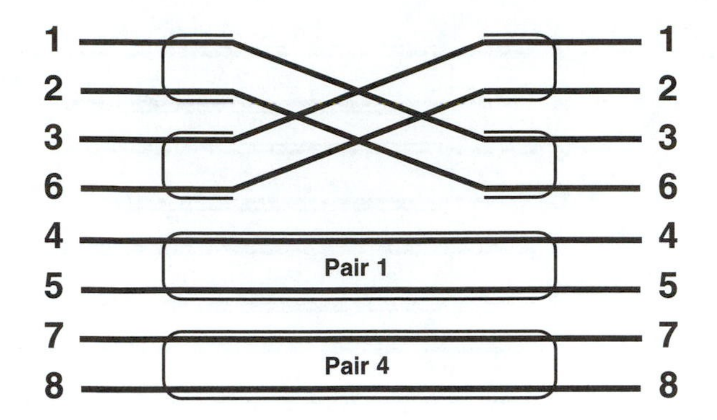


**Figure 8–7.** Crossed pairs where both conductors of two pairs are miswired.

ed. (Note that this unusual error, which requires two equal mistakes on both ends, will only be a problem with LAN connections, where crosstalk is a problem, and, since those networks should be tested with a proper tester that tests crosstalk and impedance, you will find the error during testing.)

*Conclusion.* Remember that these miswirings can occur in any combination. A common error is simply mistaking the color coding and mixing up wires. Another common error is terminating as T568A on one end and T568B or USOC on the other. The wire mapper should detect these errors.

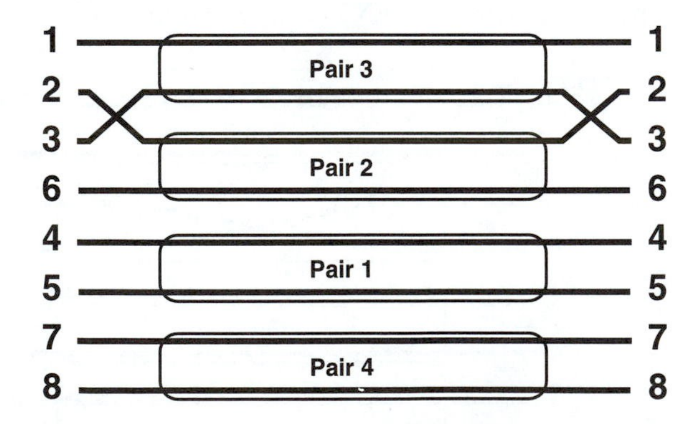


**Figure 8–8.** Split pairs have proper wire mapping, but will cause crosstalk problems.

## IMPEDANCE, RESISTANCE, AND RETURN LOSS

Impedance is the "resistance" of the cable at the frequency of signals transmitted. Return loss refers to reflections that occur at changes in impedance. These reflections can cause errors in signal transmission if they are too large. UTP cable is specified to have a nominal impedance of 100 +/- 15 ohms. For high-speed data, both impedance and return loss are functions of the signal frequency, and impedance tends to decrease with frequency.

Cables, connectors, and other hardware for high-speed networks are designed to have very consistent impedance to prevent reflections. For cable, that means having the size of the conductors, twist of the pairs, and insulation materials carefully controlled. The term *structural return loss* is used to refer to the reflections caused by variations of the impedance of the cable itself along its length. Even variations in production of cable can result in varying impedance, so consistently using the same cable in a network can minimize problems.

At connectors, the twist of each pair must be maintained to within 1/2 inch, or 13 mm, of the connection points to prevent crosstalk and undesirable reflections. If one expects consistent high-speed network performance, every component in the cable plant must be rated for Cat 5 and installed precisely.

The impedance typically goes down with higher frequency, and return loss will vary significantly with frequency. Both should therefore be measured with a tester that tests at a frequency consistent with the network planned for the cable plant. At the present time, there is no industry agreement on field-testing requirements for return loss, although this is expected to change.

Resistance is the DC component of the impedance and can be measured with a DMM. Every cable has a rated resistance in ohms/foot. If you know the length of the cable, you can test resistance with your DMM; if you know the resistance in ohms/foot, you can measure the total resistance of a cable and determine the length.

## CABLE LENGTH

Cable length needs to be known to verify that the length is within the limitations of the design standards and for future reference in moves, changes, or troubleshooting. The length of the cable can be estimated by measuring resistance as described in the preceeding text or by using a time domain reflectometer (TDR).

Time domain reflectometry works like radar, sending an electrical pulse down the cable to an open end, where the signal is reflected back to the transmitting end. By knowing the characteristic speed of the signal in the cable, called the nominal velocity of propagation (NVP), and the round-trip transit time, you can calculate the length of the cable.

The NVP is an average value; actual cable samples may vary as much as 10 percent among production lots. Even each pair, which has different twist rates to reduce crosstalk, will have a different velocity of propogation, varying as much as 3 to 4 percent, and a different physical length. Add to this the inherent inaccuracy of the instrument, another 2 to 3 percent, and you see the measurement is a good approximation of the length, but not an exact measurement. All testers will measure and report the length of each pair.

TDRs offer another valuable piece of information—they find opens, shorts, and terminations. If you see no reflection, the end of the cable is properly terminated. If your return pulse is the same polarity as the transmitted pulse, the cable is open at the end. And if the return pulse is of opposite polarity to the transmitted pulse, the cable is shorted. Thus the TDR can find cable faults by type and location, enhancing its use as a troubleshooting tool.

## ATTENUATION

As an electrical signal travels down the cable, the impedance causes attenuation (Figure 8–9). At the far end, the signal will be smaller than at the transmitter. It is important that the attenuation be less than a specified value so the received signal is of adequate strength for proper data transmission.

Attenuation is expressed in dB (decibels) where 20 dB is an attenuation factor of ten. Like everything else we have discussed, attenuation is a function of frequency, increasing at higher frequencies, and must be tested at operating frequencies. Testers will test at the frequencies specified in the category of cable being tested, up to 16 MHz for Cat 3, 20 MHz for Cat 4, and 100 MHz for Cat 5.

Measuring attenuation requires an instrument at each end of the cable, one to transmit a known value signal and one at the far end to measure the signal level and calculate the attenuation. Pass/fail criteria are set by TSB-67 and are given in the following discussion of certification.

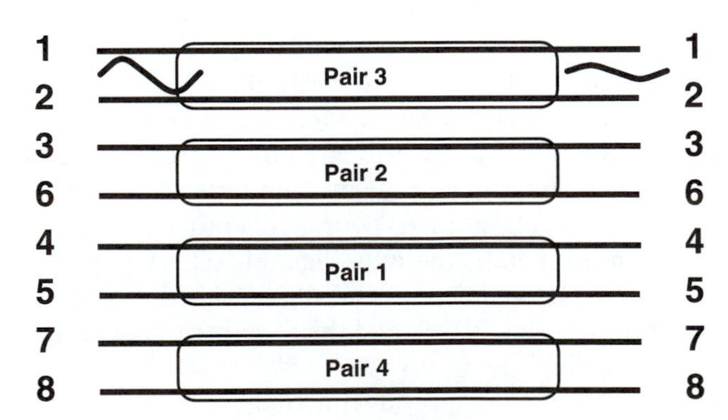


**Figure 8–9.** Attenuation in the cabling causes a lower signal amplitude at the receiver end.

## NEAR END CROSSTALK (NEXT)

In a cable that has four pairs of electrical conductors, whenever one pair is carrying signals, it may couple some of its energy into an adjacent pair (Figure 8–10). If a signal is being transmitted from the other end simultaneously, the proper signal may be compromised by interference by the crosstalk.

Each pair works like an antenna; the pair carrying a signal is the transmitter, and every other pair is a receiver. The cable construction, including the variation in twist rates in the pairs, is designed to minimize crosstalk. Like everything else, crosstalk is frequency-dependent, so it must be tested over the full frequency range specified for the category of cable being installed.

As stated earlier, at the connectors, where crosstalk is most critical, the twist of each pair must be maintained to within 1/2 inch or 13 mm of the connection points to prevent undesirable reflections. If one expects consistent high-speed network performance, every component in the cable plant must be rated for Cat 5 and installed precisely.

Testing crosstalk is quite simple. Terminate the far end of each pair to prevent reflections that would interfere with crosstalk measurements, transmit a signal on one pair, measure the coupled signal on another pair, and calculate the crosstalk in dB, just like attenuation. As with attenuation, pass/fail criteria are set by TSB-67 and are given in the following discussion on certification.

Note that each pair must be tested against all the other pairs for a total of six tests, and the test must be repeated at both ends of the cables for it to be valid.

The most common failure for NEXT is improper termination. If the twists are not maintained to within 1/2 inch (13 mm) of the termination, NEXT will fail. Poor-quality components, including patchcords, termination blocks, plugs, jacks, or cables, will cause failures also, especially at high frequencies.

As mentioned earlier, split pairs will cause unbalanced transmission and high crosstalk. It is with NEXT testing that you will likely find these problems.

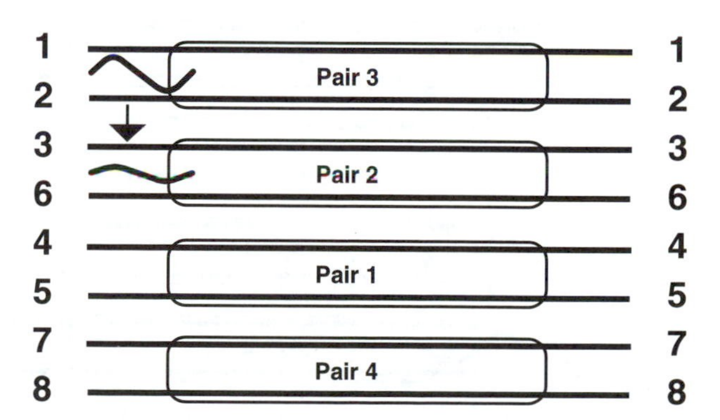


**Figure 8–10.** Near end crosstalk (NEXT) is signal coupling at the transmitter end.

## POWER SUM NEXT

Ethernet and Token Ring use only two pairs out of the four-pair UTP cable. Some of the new higher-speed networks like Fast Ethernet, ATM, and Gigabit Ethernet use all four pairs. Because they transmit over multiple pairs simultaneously, crosstalk occurs from several pairs at once. It can be a bigger problem and needs testing. Power sum NEXT measures crosstalk on one pair while all other pairs are transmitting (Figure 8–11). This test has already been implemented by several manufacturers in their Cat 5 testers but has not yet been included in TSB-67.

## ATTENUATION TO CROSSTALK RATIO (ACR)

In a duplex communication link, like networks running on UTP cabling, signals can be traveling in both directions simultaneously. Thus, at the receiver end of the cable, one can have a signal from the other end attenuated by the cable, and crosstalk coupled from the local transmitter, simultaneously arriving at the receiver. Since both ends transmit signals of approximately the same amplitude, we want the received signal to be higher than the interfering crosstalk.

Attenuation to crosstalk ratio (ACR) is the measure of this situation (Figure 8–12) and an excellent indication of the overall quality of the cable link. For example, if the crosstalk is 35 dB and the attenuation is 15 dB, our received signal will be 20 dB larger (a factor of ten) than the crosstalk. If the crosstalk is 28 dB and the attenuation is 22 dB, our received signal will be 6 dB larger (a factor of two) than the crosstalk. Thus the higher the ACR, the better the cabling performance.

## PROPAGATION DELAY AND DELAY SKEW

Propagation delay (or simply delay) is a measure of the time it takes for an electrical signal (traveling at about two-thirds the speed of light) to reach the far end of

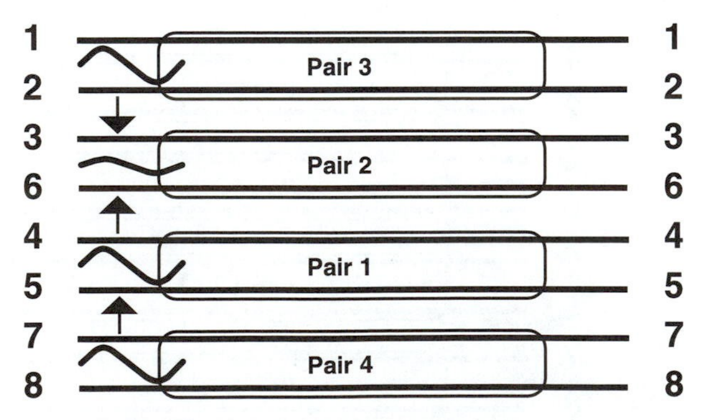


**Figure 8–11.** Power sum NEXT measures the coupling effect of all other pairs into one of the four pairs in the cable.

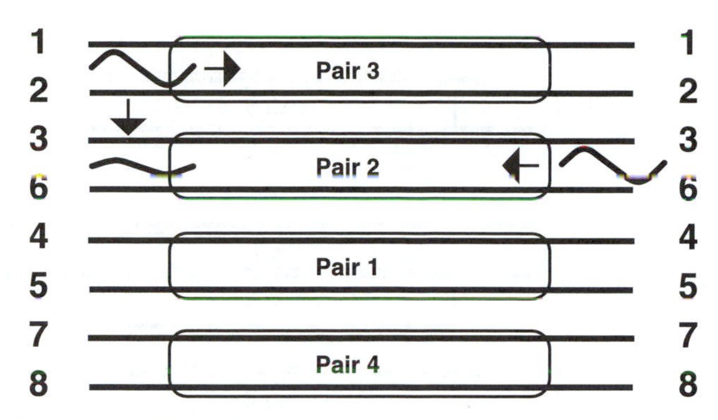


**Figure 8–12.** ACR is the difference between an attenuated signal at the receiver and crosstalk from an adjacent transmitter.

the cable. Testers measure delay as part of the process of measuring length. Since each pair has a different length due to the twist rate and a different NVP, there will be variations among the four pairs. This is not a problem with most networks, as the signals are carried on only one pair in each direction. However, many of the high-speed networks (100 MB/s and higher) are using two or even four pairs in each direction, so if the parallel signals arrive too far apart, data errors will occur.

The design of Cat 5 cable includes having a different twist rate on each pair of wires; a more tightly twisted pair will have slightly longer wires for a given cable length. Thus, a signal will take different times for end-to-end transmission for each pair, and NVP can also vary with wire insulation, also causing differences in transit times. Some cables have been built with different (cheaper) insulation on the pairs not normally used with Ethernet or Token Ring, and the variation in transit times may be more than is tolerable.

The maximum difference in transit times between all four pairs is called "delay skew" (Figure 8–13). The delay skew must be less than a specified amount to allow the Cat 5 cable to work with the high-speed networks. Since delay skew is a function of the cable itself, it is important to buy only Cat 5 cable that meets the required specifications as outlined in EIA/TIA 568A. However, delay skew is not likely to be a field-test requirement, since it cannot be easily affected by installation practices.

You may still find cable that does not meet this new specification in existing installations, and you may need to verify that skew specifications are met. A big problem was created in the past few years during the shortage of the plenum-rated Teflon wire insulation. A fair amount of Cat 5 cable was installed with pairs having different insulation materials and therefore widely varying NVP.

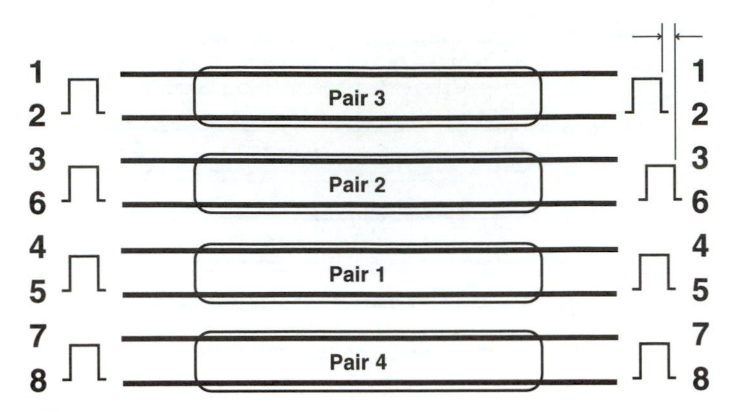


**Figure 8–13.** Delay skew measures the timing differences between pairs.

## CABLE PLANT "CERTIFICATION" TO STANDARDS

The industry has agreed on a set of test standards for UTP cabling technical service bulletin TSB67 of EIA/TIA 568. Of all the tests described so far, only wire mapping, length, NEXT, and attenuation are required at this time. Some of the automatic testers will give you additional data, like ACR or delay skew. They will also give you a "pass/fail" result, and perhaps even an indication if the result is uncertain, as when the measured result is closer to the limit than the actual instrument accuracy.

Length testing requires the link be less than 94 meters and the channel to be less than 100 meters, including the test equipment patchcords. Wire mapping to TSB-67 requires connections per 568A or 568B. Other connections, including USOC and DECconnect, while perhaps valid for the installation, will not meet TSB-67.

## NEXT AND ATTENUATION

EIA/TIA TSB-67 specifications for NEXT and attenuation are shown in the graphs in Figures 8–14 and 8–15. There are two different specifications to accommodate the installed cabling with or without final patchcords used to connect the equipment to the cabling.

A "channel" is the "end-to-end" connection from the equipment on either end, including all the patchcords. A "basic link" includes the wiring from the local outlet to the final termination in the telecom closet. The channel performance specifications are worse than links, as they include two additional connections on either end.

From the graphs in Figures 8–14 and 8–15, you can see ACR as it has been described in the preceding text. As attenuation and NEXT increase, the ACR or signal-to-noise ratio degrades, shown by the converging curves.

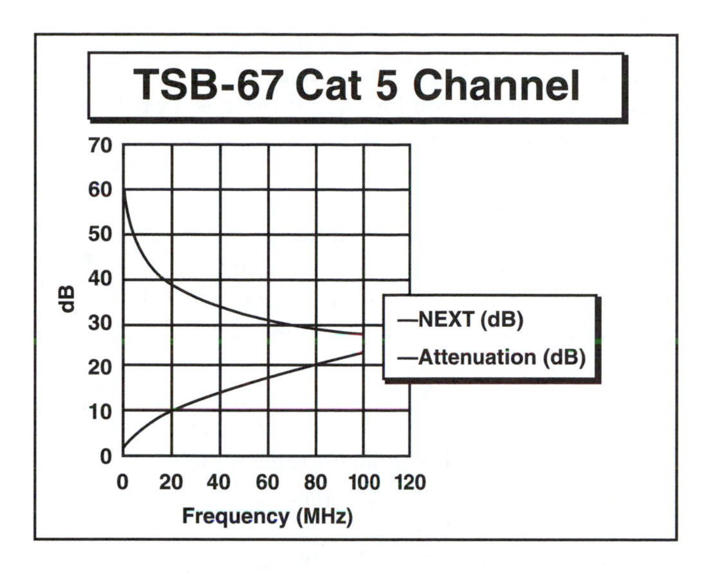


**Figure 8–14.** Attenuation and NEXT end-to-end in a Cat 5 channel.

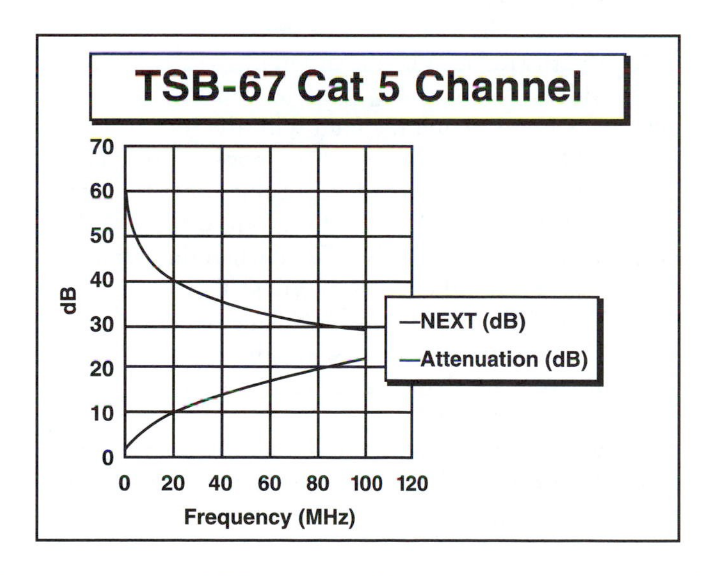


**Figure 8–15.** Attenuation and NEXT end-to-end in a Cat 5 link removes the effect of interconnecting patch cords on either end.

## CHAPTER REVIEW

1. What is used as a simple coax tester?
2. What should you look for if there is a problem with a coax cable transmission?
3. What UTP test instruments do you use to test low-speed wire or Cat 5 high-speed wire for correct connections?
4. Why is Cat 5 a better choice to install than Cat 3?
5. What is used to test wire mapping, length, attenuation, and NEXT?
6. What are the colors used in the 4-pair UTP cable? List the pairs together.
7. Other than the EIA/TIA 568A or EIA/TIA 568B, what variations are used but not recommended for structured cabling?
8. What company's standard established T568B wiring?
9. What is most important in any cable?
10. What does a time domain reflectometer (TDR) test?
11. What happens in a split pair?
12. What is impedance?
13. What are reflections that occur at changes in impedance?
14. What is it when the cable has a reflection from a variation in impedance in the cable?
15. How much untwisting is allowable before you lose performance in Cat 5 cable?
16. What do you need to know to use a time domain reflectometer (TDR) to calculate the length of a cable?
17. What is reduced by the different twist rates in the separate pairs of UTP cable?
18. *Matching:* Match the fault with the return pulse indication on a TDR.
    A. Open at the end — (1) Opposite polarity to the transmitted pulse
    B. Properly terminated — (2) Same polarity as the transmitted pulse
    C. Shorted — (3) No return pulse
19. Near end crosstalk (NEXT) is when the transmitting pair couples part of its signal to other pairs of wires. T or F
20. What does power sum NEXT measure?
21. What occurs in a delay skew and propagation delay?
22. Different insulation in a Cat 5 wire does not affect the performance of the wire. T or F
23. What are the specifications for a cable plant length for a link and a channel?
24. What does a channel include that a link does not include when dealing with length?

CHAPTER

# 9

# WIRING TERMINATION PRACTICES

The purpose of this chapter is to show you how to handle and terminate cable for telephone, LAN, and CATV/CCTV video networks. All the tools and materials needed are described in detail to allow you to follow these directions and learn how to properly install and terminate cables.

To get the most out of the exercises in this chapter, you should already have familiarized yourself with data, voice, and video wiring by having studied the earlier sections of this book. This chapter covers both coaxial cable for CATV/CCTV and Category 5 wiring for local area networks (LANs).

## COAX WIRING

### Coax CATV Cable

In this section, we will concentrate on installing connectors, the most critical part of coax cable installation. All CATV connections are made with type F connectors, shown in Figure 9–1 along with a BNC bayonet connector for comparison. The F connector is a screw-on or crimp coax connector that uses the center conductor of the cable as the contact and that crimps or screws on the cable jacket and shielding to make contact with the connector body.

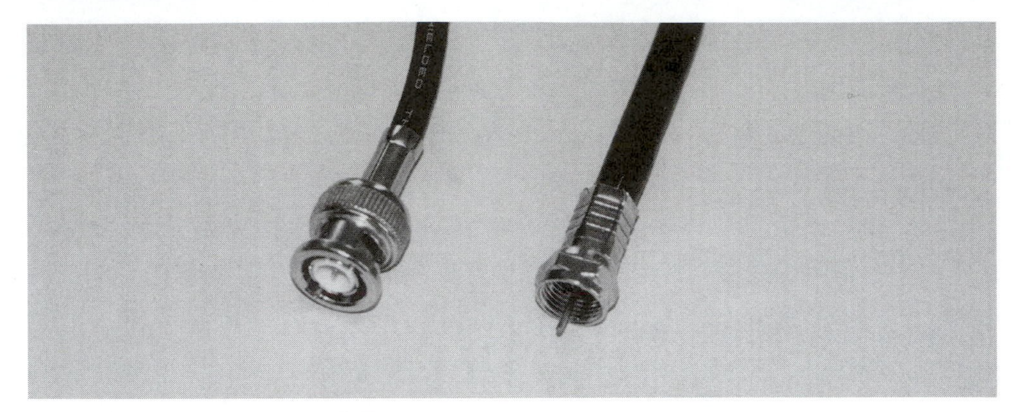

**Figure 9–1.** Coaxial cable connectors. BNC on the left, Type F on the right.

Screw-on F connectors are often used to make up cables to go from CATV wall outlets to the TV, but the quality of these connectors may allow signal leakage or reflection. They are not recommended for usage in permanent wiring; crimp-on connectors should be used instead.

The most common cable for CATV premises wiring is a coaxial cable known as RG-6 (Figure 9–2). It has a large center conductor surrounded by a plastic foam dielectric that is covered by a metal foil, then a woven braid for the shield, and finally a jacket of insulating material appropriate for the locale of the installation.

### Tools

For the exercises in this chapter, you will need a coax cable stripper and the crimp tool with the proper coax connector jaws installed.

**Coaxial Cable Construction**

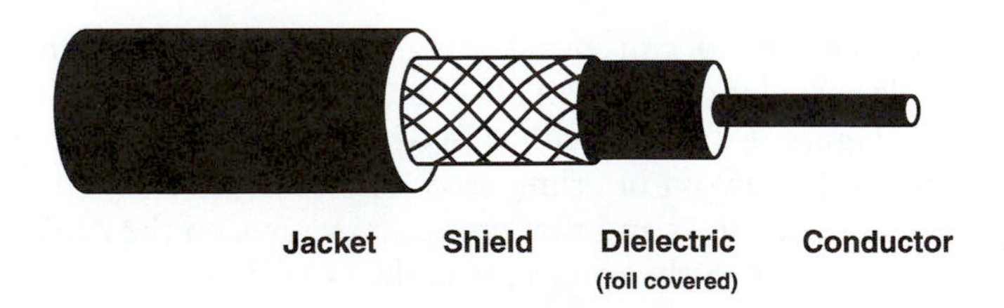


**Figure 9–2.** RG-6 coaxial cable construction.

*Coax strippers.* The coax stripper (Figure 9–3) prepares the cable for termination by cutting all the outer layers of the cable off to expose the center conductor and then removing the jacket to allow the connector body to make contact with the shield. Several types of strippers exist. One is an adjustable stripper that can make several simultaneous cuts to prepare a coax cable for termination in one step. This cutter must be adjusted for the cable type being stripped. If the cable is not being stripped correctly, refer to the instruction manual for the stripper information and adjust the cutting blades accordingly. Other types of strippers cut the cable differently, depending on the direction of the cut. These types are less expensive and easier to adjust and use, although they take more time to strip the cable.

*Crimpers.* The crimper clamps the connector to the cable permanently. Most crimpers have hexagonal crimps sized for each connector type and replaceable jaws for different connectors. A unique set of jaws is needed for type F coax connectors, BNC connectors, and Cat 5 8-pin modular connectors. Make sure the proper jaws are installed before using the crimper.

Most crimpers are ratcheted, so that once a crimp is started, you must complete the cycle to ensure the crimp is properly made. These crimpers usually have a release lever to bypass the ratcheting, so it is a good idea to find the lever and make certain you know how it works in case you need to abort a crimp.

### Examining Cable Construction

Using the coax cable stripper, insert the coax cable into the stripper and rotate the stripper several turns to make the cuts. Pull the cable from the stripper, and it should be ready for termination. (Figure 9–4)

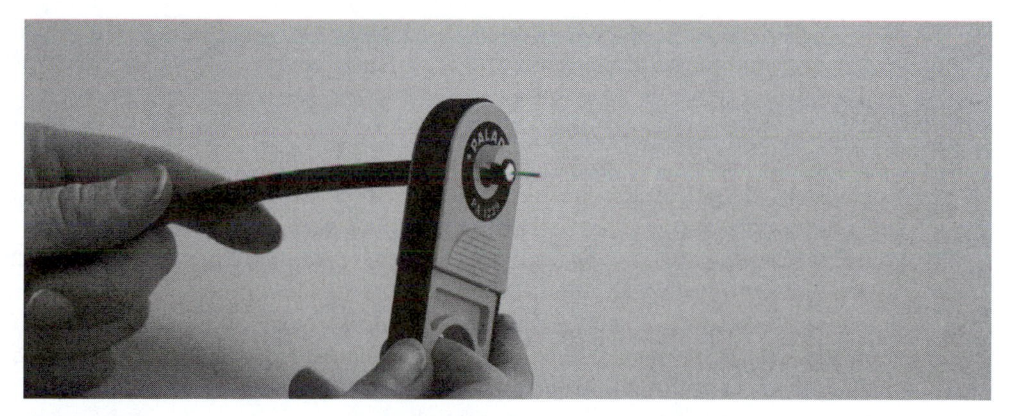

**Figure 9–3.** RG-6 coax cable for termination.

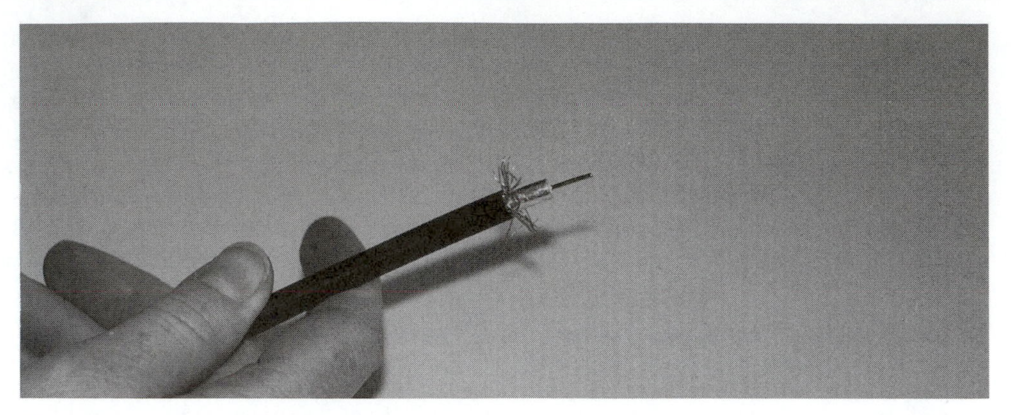

**Figure 9–4.** RG-6 cable ready for termination

Examine the cable carefully. Note the center conductor that also acts as the center pin of the connector. See how the foam dielectric separates the center conductor from the shield and how it is covered with a foil layer that also assists in shielding the cable. Note the woven shield over the foam. It should be tightly woven to prevent signal leakage.

### Terminating Coax with F Connectors

The process of terminating these cables is as follows:

1. Use the coax stripper on the cable to strip the cable for termination. The outer insulating jacket must be stripped away, exposing the braided shield.
2. The braided shield must be pulled back, over the outer jacket, leaving the inner insulation and its foil shield exposed.
3. The exposed center conductor should be checked to make sure it is clean.
4. The connector is then placed on the end of the cable and crimped or tightened down (Figure 9–5).

### Comments on Installation

Installing video cabling is relatively simple. Coax cables must be installed with care; they may not be pulled beyond their tension limits, and may they not be sharply bent. In addition, they must remain safe from physical damage and from environmental hazards. Care must also be taken when strapping or (especially) stapling cables to structural surfaces (walls, ceilings, and so on). Special staples are available for coax cables. If the staple or strap is cinched too tightly, it will deform the cable and alter its transmission characteristics. If the staple or strap is overly tight, the system may not work properly, or even at all.

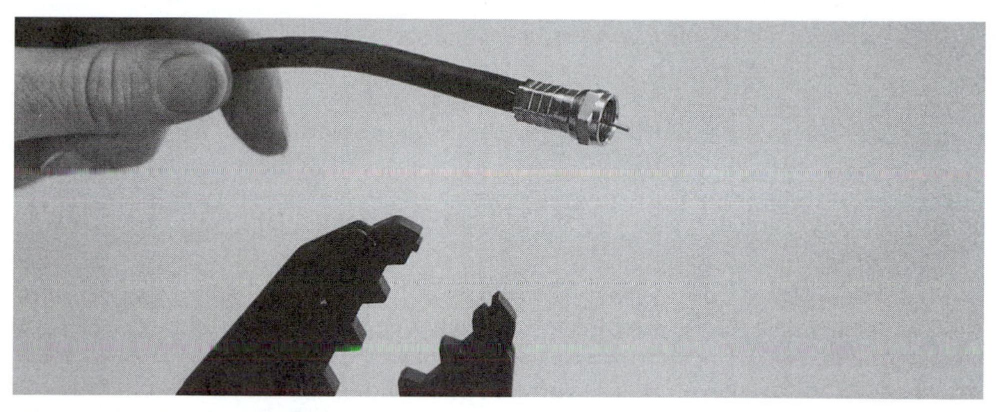

**Figure 9–5.** Finished F connector after crimping.

Perhaps most important with CATV cables is proper cable selection and termination. CATV cables are directly connected to the public CATV network. FCC rules limit signal leakage, so it is important to use good cable with proper shielding and to terminate properly in order to prevent signal interference with other electronic devices. In addition, poor termination can cause reflections in the cable that affect the return path, or connection back to the system. Since more networks are upgrading to a greater number of channels of programming and use, or are planning to use, cable modems for Internet connections, proper CATV installation has become more important

## CATEGORY 3 AND 5 STRUCTURED CABLING

### Introduction

Most commercial installations use Category 5 wiring for LANs and Category 3 or Category 5 for telephones. Most new or rehabilitated construction will install the cabling closely following EIA/TIA 568 with two UTP cables per work area. Sometimes a Cat 5 cable is used for LANs and a Cat 3 cable is used for telephones, but today, often two Cat 5 cables are used, since the cost of Cat 5 is little more than that of Cat 3, and using only Cat 5 cables facilitates future upgrades and simplifies managing cable plants.

In this section we show how to install a Cat 5 cable for LANs and a Cat 3 cable for telephones, using standard hardware, exactly as you would in a real installation.

### Warning!

Category 5 UTP cable and all the hardware associated with it are designed for LANs operating up to 100 million bits per second (MB/s). While most of today's LANs are operating at 10 MB/s, most users assume that, at some future date, they will upgrade to the higher bit rates. Therefore it is necessary to install the network cabling properly and test to Cat 5 specifications. Most users will specify operation to Cat 5 specifications and require testing data to verify performance.

In order to ensure Cat 5 performance throughout the network cabling, it is very important to use only Cat 5 hardware, including termination blocks, jacks, and plugs, and to follow installation procedures rigorously. Sloppy installation practices will degrade Cat 5 components to Cat 3 specifications. Numerous articles in the trade press have said that as many as 80 percent of all Cat 5 installations will not meet Cat 5 performance specifications due to poor workmanship. Learn how to install Cat 5 properly so you do not add to that statistic!

The key to Cat 5 performance is maintaining balanced transmission over the controlled twist of the wires. At every connection, it is mandatory to maintain the twist as close to the connections as possible, never allowing the pairs to untwist more than 1/2 inch (13 mm). In addition, the twist can be damaged by pulling the cable too hard. The specification for pulling tension is only 25 pounds. As part of this exercise, you will pull cable with a scale to see how little tension this is! And it is mandatory to never kink the cable when pulling it. Feed it carefully and avoid sharp corners when pulling.

### Examining Cable Construction

Category 3 and Category 5 cables (Figure 9–6) are made from four twisted pairs of 24- to 26-gauge copper wire. LAN cabling is solid wire, but patchcords are usually made of stranded wire for greater flexibility. Each pair is twisted at a different rate and is color-coded.

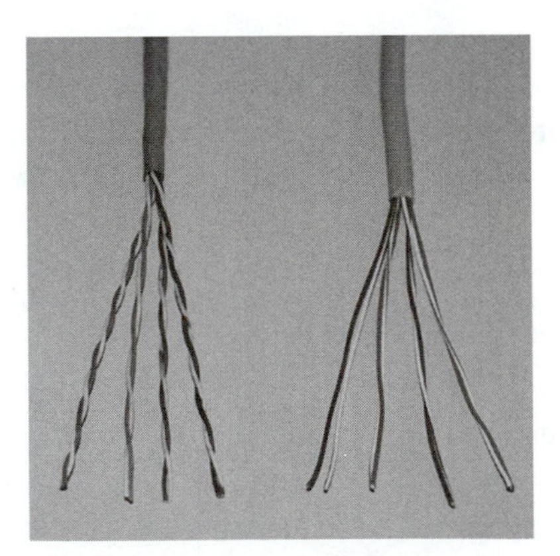

**Figure 9–6.** Cat 5 cable (L) has tighter twists in the pair than Cat 3 cable (R).

Take a sample of Cat 5 cable and strip off about 2 to 4 inches of the outer jacket. Spread out the four pairs and note the color coding and differences in twist rates in each pair. If you look closely, you can see the differences in twist rates. Now strip off about 2 to 4 inches of the jacket of Category 3 cable and examine the pairs. Note how little twist is present in the Cat 3 pairs. The main difference in performance of Cat 3 and Cat 5 cable is the tightness and consistency of the twists.

To terminate these cables, you will need to separate the wires and order them according to the proper color code for the type of connection being made (T568A or T568B). In addition, it is imperative to maintain the twist in each pair of the Cat 5 cable to within 1/2 inch (13 mm) of the termination point. Cat 3 cable can have as much as an inch of untwist and still work with 10BaseT networks, but maintaining the twists as closely as possible is still recommended.

### Pulling Cat 5 Cable

It is important to not stress the cable more than necessary when pulling. The maximum pulling tension on Cat 5 cable is only 25 pounds. The performance of Cat 5 cable is determined by the careful construction of the twists in the pairs. Excess tension on the cable may change the twists in the cables, causing an increase in crosstalk, attenuation or both.

A tension of 25 pounds is not very much. Sometimes cable is installed in conduit where the pulling tension can get quite high if there are significant bends in the conduit run. Other times, cable is run in open spaces but has to work its way around obstacles that may create tension or, even worse, try to kink the cable.

To illustrate how much force is involved in tension of 25 pounds, simulate a cable pull, but with a scale attached to measure the pulling tension. A fishing scale works well, although it should have a range greater than 25 pounds. Find a sturdy place to attach one end of the cable, then set up the exercise shown in Figure 9–7.

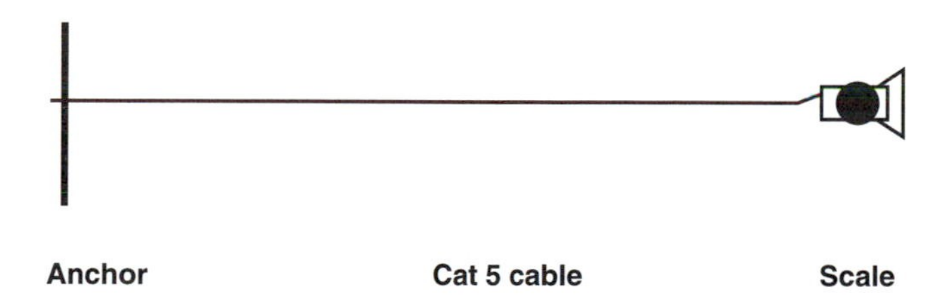


**Figure 9–7.** Pulling exercise for Cat 5 to "feel" pulling tension.

Cut a 6-foot (about 2-meter) section of Cat 5 cable from the roll. Attach one end of it to a sturdy place that will withstand a hefty pull. Attach the other end to the pulling rope that includes the scale. First, pull as hard as you think you should and, while holding the tension, check the scale. Are you over or under 25 pounds? Now pull while watching the scale to see what effort creates 25 pounds of tension. Try to remember this effort—and do not exceed it!

### Installing a Simulated Cat 5 Network

The EIA/TIA 568 standard has defined the horizontal connection as up to 90 meters of permanently installed cable and up to 10 meters of patchcords. In this exercise, you will use simulate the installation of a link. The installation will include wiring two jacks and one 110 punchdown block in the link, with a total length of almost 100 meters. You will also make two patchcords. When you finish, the link will look like that shown in Figure 9–8.

### The Wire U® Training Board

In teaching the Wire U® program, we build a training board on plywood like the one shown in Figure 9–9. The training board has the two most common punchdown blocks, a 110 block for Cat 5 and a 66 block for Cat 3. The 66 block is rarely used for Cat 5, but directions are provided for using it with Cat 5 if desired.

The 66 block and the Cat 3 jacks on the board are to be wired for voice, and the 110 block and Cat 5 jacks are used for wiring the LAN cable. You will need

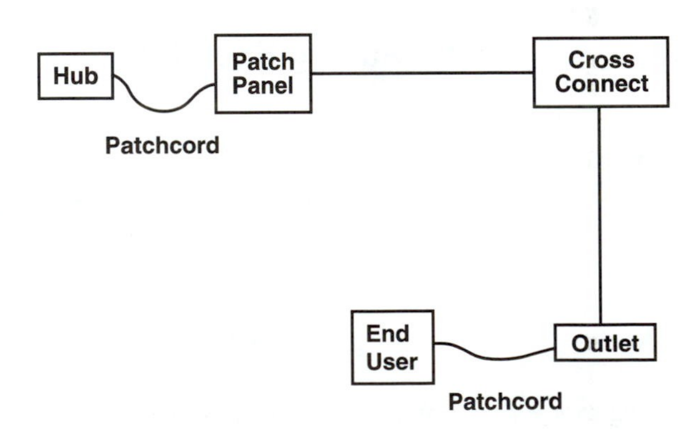


**Figure 9–8.** Simulated Cat 5 network for installation practice.

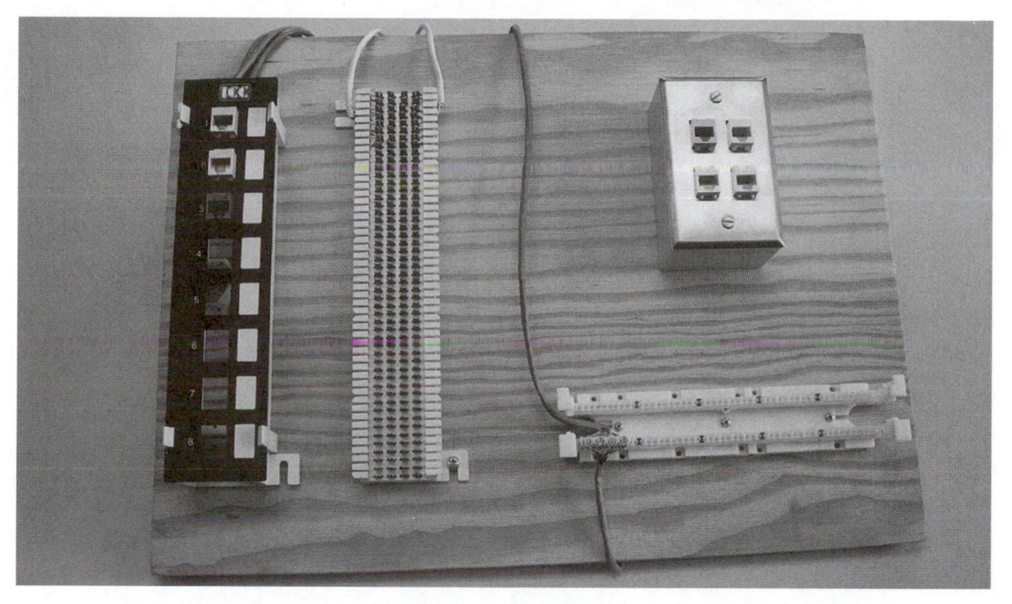


**Figure 9–9.** The Wire U® training board

about 90 meters of Cat 5 cable for the LAN cable exercise to simulate the maximum length of a 568 link. You will also need a shorter length of Cat 3 cable for telephone wiring.

We suggest you "install" the Cat 5 link by doing the punchdown at the cross-connect first, then terminating the jacks. This creates what is called a link and should be tested first. Use a wire mapper to check correct connections through the link. If you have access to a Cat 5 tester, use it to test the link for dynamic performance to 100 MHz also.

After testing the link, make two patchcords about 3 meters (10 feet) long. Attaching them to the jacks completes the *channel.* Retest for wire map and Cat 5 performance, as you did in testing the link.

### Terminating Cat 5 Cable

The proper termination of Cat 5 cable is mandatory to maintain the performance specified by EIA/TIA 568 standards and expected by your customers. As mentioned earlier, proper cable pulling and handling during installation is important to maintain cable performance. But most problems with Cat 5 installations occur at terminations. If you do not maintain the twists right up to the terminations or do not properly terminate the IDC (insulation displacement connector) connections, the performance of the cable will be compromised. Remember, you must use only Cat 5 rated components for every part of the network cabling!

The majority of Cat 5 terminations are made with standard IDC connections, using 110 punchdowns or crimped RJ-45 connectors. Almost every manufacturer now has alternative design plugs, jacks, and patch panels that do not require tools. Each of these has its own termination procedure that, while similar to the normal terminations, requires following the manufacturer's instructions exactly.

For the purposes of this course, we will stick to industry standard 110 blocks and crimp connectors. Once you learn these, you can easily learn how to use any other type of termination.

## Tools

*Cable jacket stripper.* This lightweight stripper (Figure 9–10) will make cuts on the outside of UTP cable jackets to allow removal of the required length of jacket quickly and easily. Several other types are available, including very simple ones that are often given away by manufacturers as promotional items. It is very important to not nick the wires when cutting the jacket, so try to just score the jacket, twist it to break it, and pull it off. You can also cut off an inch or so of the jacket and use the rip cord to slit the jacket as far as needed, then cut it off.

*Punchdown tool with 110 blade.* This impact punchdown tool (Figure 9–11) inserts the wire into the 110 block and cuts the ends off. It can be used with solid or stranded 28 to 22 AWG single insulation wire. The blades provide "punch and cut," or just "punch" by flipping the blade over. The extra blade is usually stored in the bot-

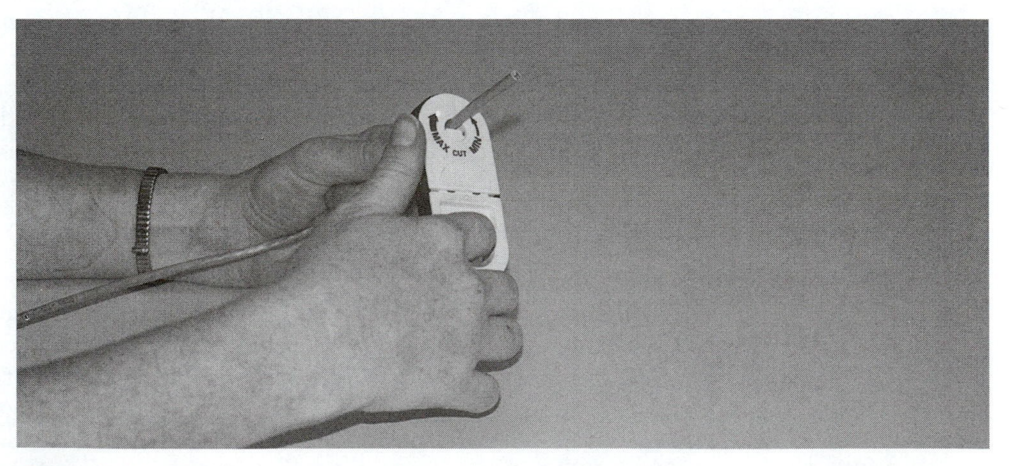


**Figure 9–10.** Remove the cable jacket with a cable stripper, using care not to nick the wire.

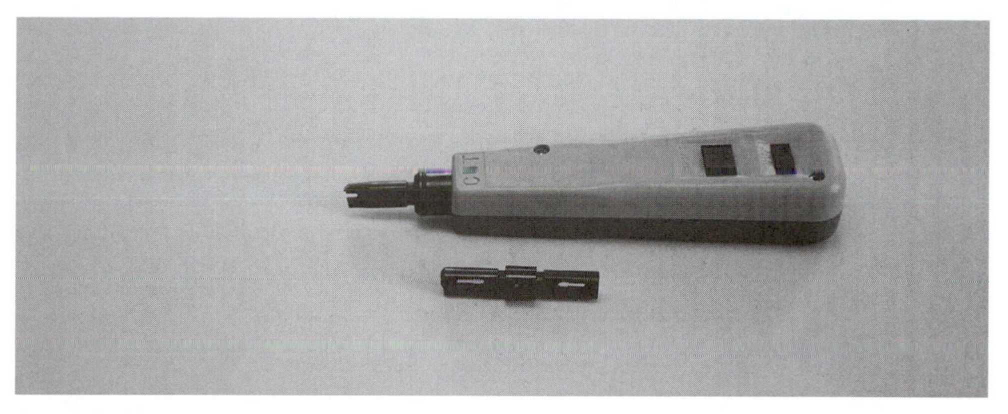

**Figure 9–11.** Punchdown tool with 110 blade (installed) and 66 blade.

tom of the handle. Impact spring compression is adjustable to high or low with the dial in the handle.

Select the 110 blade and insert it in the tool with the "cut" end out, for performing both punchdown and a cut. Install the blade into the top of the tool by pulling down the outside retaining clip and inserting the blade. Line up the key located on the side of the blade with the internal socket. Release the retaining clip to lock the blade into position.

*Crimp tool.* Most crimpers have replaceable jaws for different connectors (Figure 9–12). Jaws are available for most connectors, including 4-, 6- and 8-pin modular connectors and coax connectors. Make sure the proper jaws are installed before beginning the exercise.

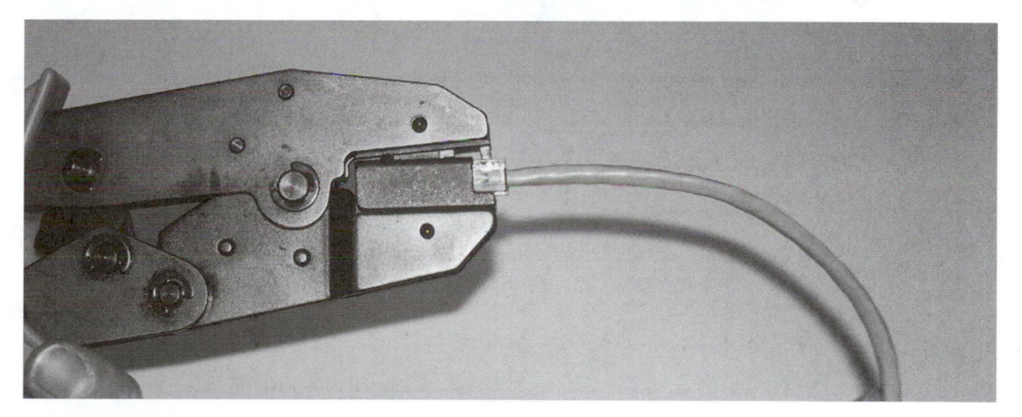

**Figure 9–12.** Crimp tool with modular connectors.

### Terminating with 110 Punchdown Blocks

The 110 block is widely used as an interconnect for data wiring. It consists of several parts; which parts are used depends on the interconnection scheme used. The *wiring block* positions the wires to be terminated by holding them in a plastic base after punchdown. Some types of 110 blocks require an additional mounting piece to mount the wiring block on a panel and may help hold cables neatly in place. In use, the wires are punched down onto wiring blocks, the connecting block is snapped on, then another cable can be punched down onto the connecting block or cables with patch plugs can be used to interconnect the proper wires.

#### *Exercise*

Tools required: Jacket stripper and punchdown tool with 110 blade (cut side)

1. Strip about 2 inches of jacket off the cable using the jacket stripper.
2. Separate the pairs (but *do not untwist*) in the order shown in Figure 9–13.
3. For each pair, untwist just enough wire to place each wire in the punchdown slot. Remember, it must be untwisted less than 1/2 inch after termination.
4. Place the wires in the slots on the wiring block in color-coded order: blue, orange, green, brown, with the white wire of each pair first (Figure 9–14). Some installers remember the order of the colored pairs as BLOG

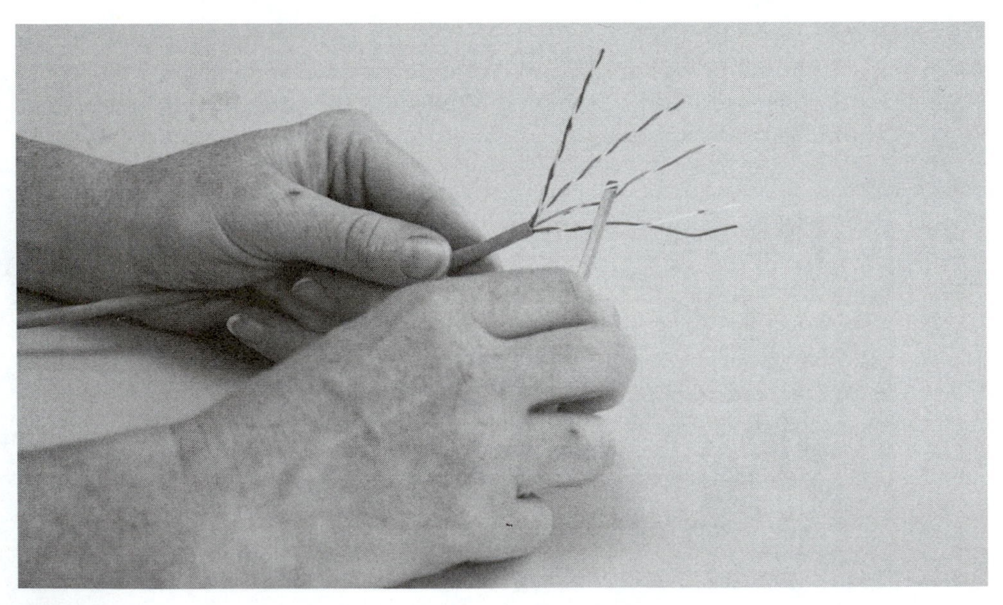

**Figure 9–13.** Pairs can be untwisted with a small screwdriver.

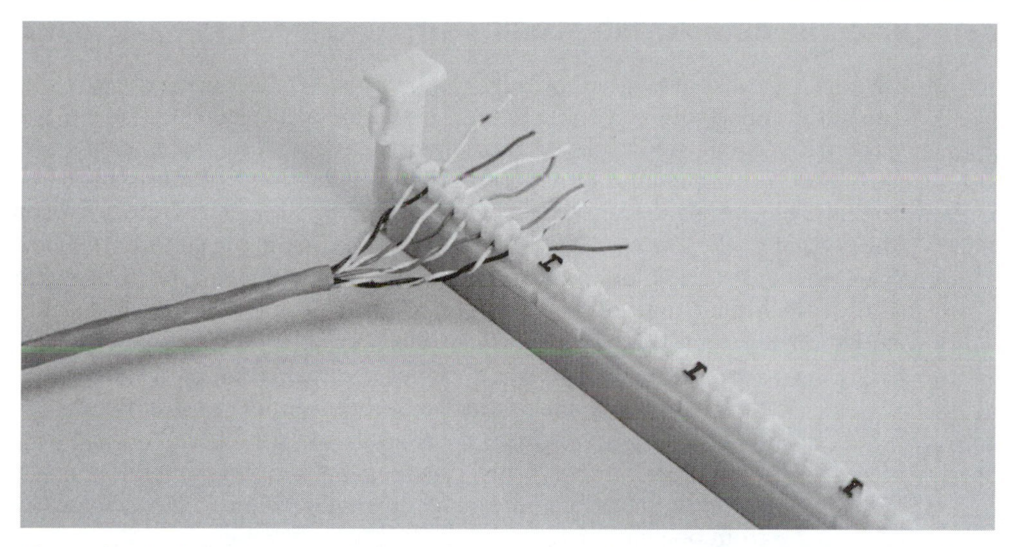

**Figure 9–14.** Place the wires in the wiring block, maintaining twists as close as possible.

for Blue-Orange-Green (with Brown following by default), but some 110 blocks have color coding on the block itself. Remember, the white wire of the pair always goes first.

5. Punchdown with the tool, with the "cut" side of the blade on the side where the end of the wire exits the 110 block (Figure 9–15).

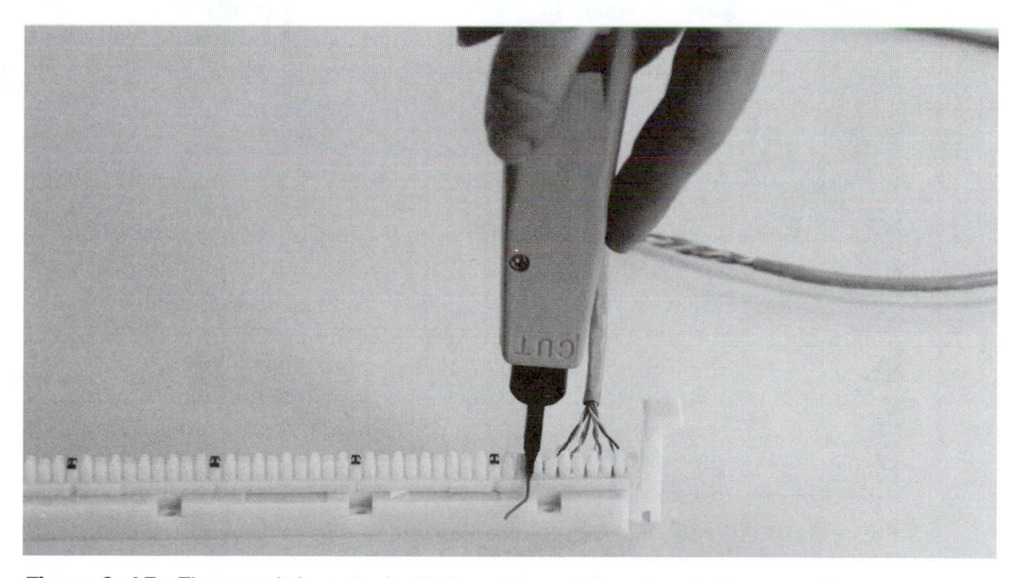


**Figure 9–15.** The punchdown tool with insert to cut the wires in the operation.

## ATTACHING THE CONNECTING BLOCK

You have probably noticed the 110 block has no connector contacts. The contacts are in the connecting block. There is an insulation displacement connector (IDC) for each wire on one side of the connecting block that inserts into the 110 block after all the wires are punched down. On the other side, the connecting block looks like the 110 block, but this time there are IDC contacts inside the block to connect to the second cable you wish to terminate with the first cable already punched down. The block also has color coding to prompt you as to which pair goes where.

Attach the connecting block to the 110 block you just punched wires into as shown in Figure 9–16.

1. Position the block with the color coding matching the wire pairs and so the IDC connectors are in the proper location over the wires in the 110 block.
2. Push the connector firmly down until seated fully and all wires are properly terminated. Use the punchdown tool to help seat the block, working from one end first.

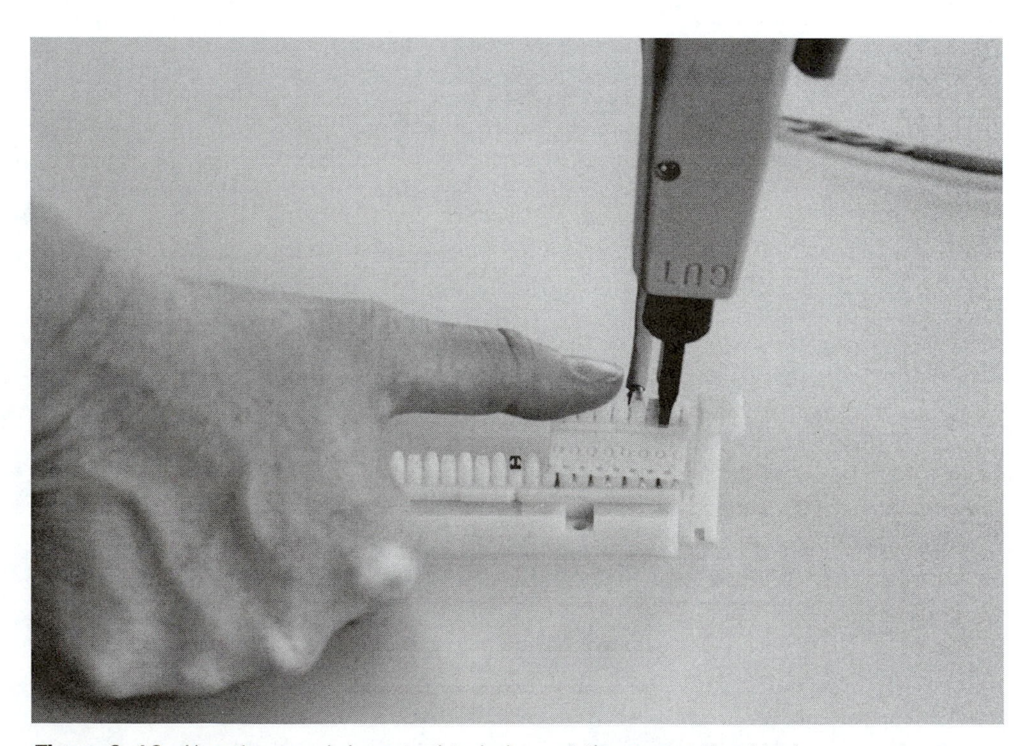

**Figure 9–16.** Use the punchdown tool to help seat the connecting block.

## CONNECTING THE MATING CABLE

We have two options in connecting another cable to the one already terminated. We can punchdown the cable to the top of the connecting block, or we can terminate the cable in a patch plug. The punchdown method is simpler, but permanent. If a move is expected, the additional cost of a patch plug on the mating cable may be a good investment.

For this exercise, we will punch down the mating cable to the top of the connecting block. Start by finding the end of the cable you want to terminate.

## CONNECTING TO THE CONNECTING BLOCK

1. Strip about 2 inches of jacket off the cable using the jacket stripper.
2. Separate the pairs (but *do not untwist*) in the same order as before.
3. For each pair, untwist just enough wire to place each wire in the punchdown slot. Remember. it must be untwisted less than 1/2 inch after termination.
4. Place each wire in order into the slot of the 110 block (Figure 9.17).
5. Punchdown with the tool, with the "cut" side of the blade on the side where the end of the wire exits the 110 block (Figure 9–18).

If we have done this properly, we will have a proper termination, with all wires correctly connected and with less than 1/2 inch of each pair untwisted. It should meet Cat 5 NEXT specifications.

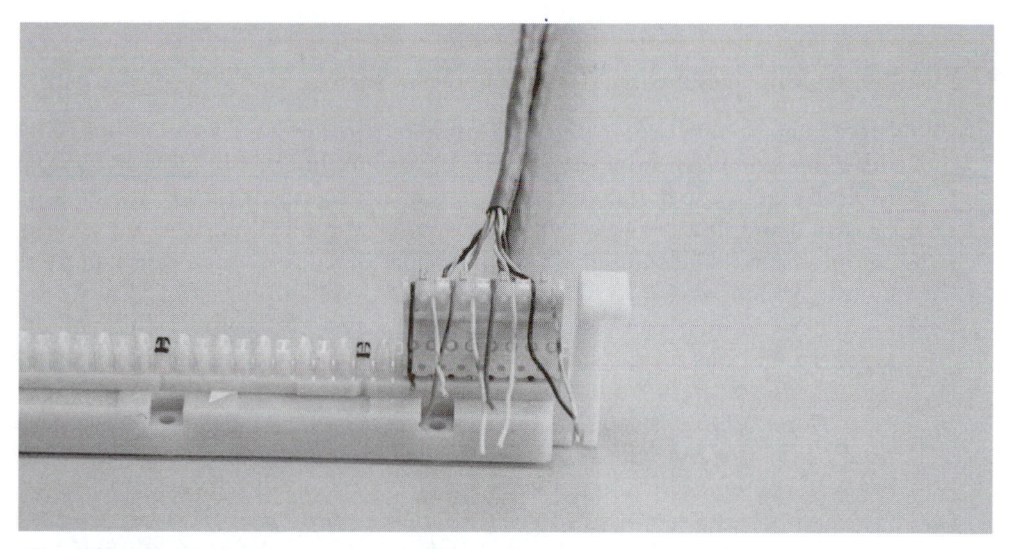

**Figure 9–17.** Wires placed properly in the connecting block.

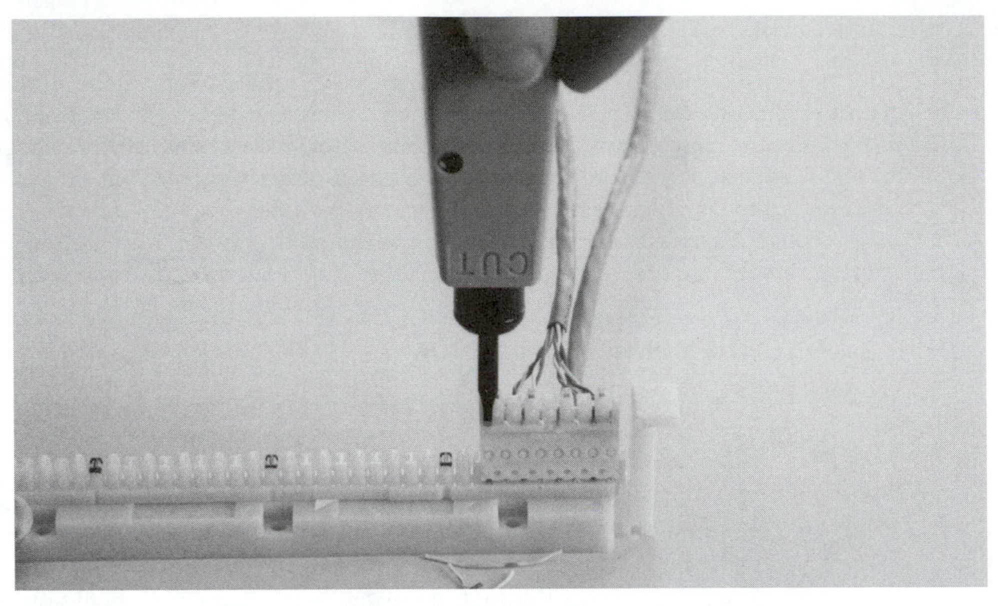


**Figure 9–18.** Punchdown the wires into the connecting block.

Once you have visually inspected your connections to make sure all wires are terminated to the same color-coded wire in the other cable, you will have finished this exercise.

### Terminating Jacks with 110 Punchdowns

Modular jacks are used with Cat 5 cable to provide for a termination at the end of the cable for connection at the work area or for patching in a telecom closet. Modular jacks use IDC connections, terminating with either 110 punchdown blocks or with a simple cut-and-crimp operation using molded parts of the jack itself. The cut-and-crimp styles are made by many manufacturers, each one slightly different, but everyone makes 110 type terminations, so we will focus on those.

The modular jacks used in the illustrations are typical jacks with 110 block type terminations for the wires. Now that you know how to use the punchdown tool on 110 blocks, we will use the same tool to terminate the cable at the jack.

***Note:*** *While the jacks all will terminate in either T568A or T568B terminations (and USOC for that matter), generally there is color coding on the back of the jack to show you how to order the wires for termination. BE CAREFUL! The jacks often include both 568A and 568B color coding (and the print is small), so be sure you use only the correct coding for that jack and do not get confused halfway through the job. A piece of narrow black drafting tape can be used to cover the color code for the wiring standard you are not using to prevent confusion. Also, since there are* *often "twists" inside the jack to reduce NEXT, the order of the wires on the back of the jack is sometimes different from the normal order for an RJ-45 plug. Make sure you know you have the wires correctly ordered according to the jack requirements!*

### Exercise

Tools required: Jacket stripper, cable cutter or crimp tool and punchdown tool with 110 blade (*cut* side), small screwdriver (optional: hand holder for jack)

1. Strip off about 2 inches of the jacket.
2. For each pair, untwist just enough wire to place each wire in the punch-down slot. Remember, it must be untwisted less than 1/2 inch.
3. Place the jack in the special fixture (Figure 9–19) to hold it securely.
4. Punchdown the wire to terminate it, and cut off the excess wire.
5. Repeat this procedure for all other wires.
6. Snap on the protective covers provided.

Once you have terminated one end of the cable, insert the jack into the outlet and snap it into place. Then repeat the process to terminate the other end of the cable, and snap that jack into its outlet.

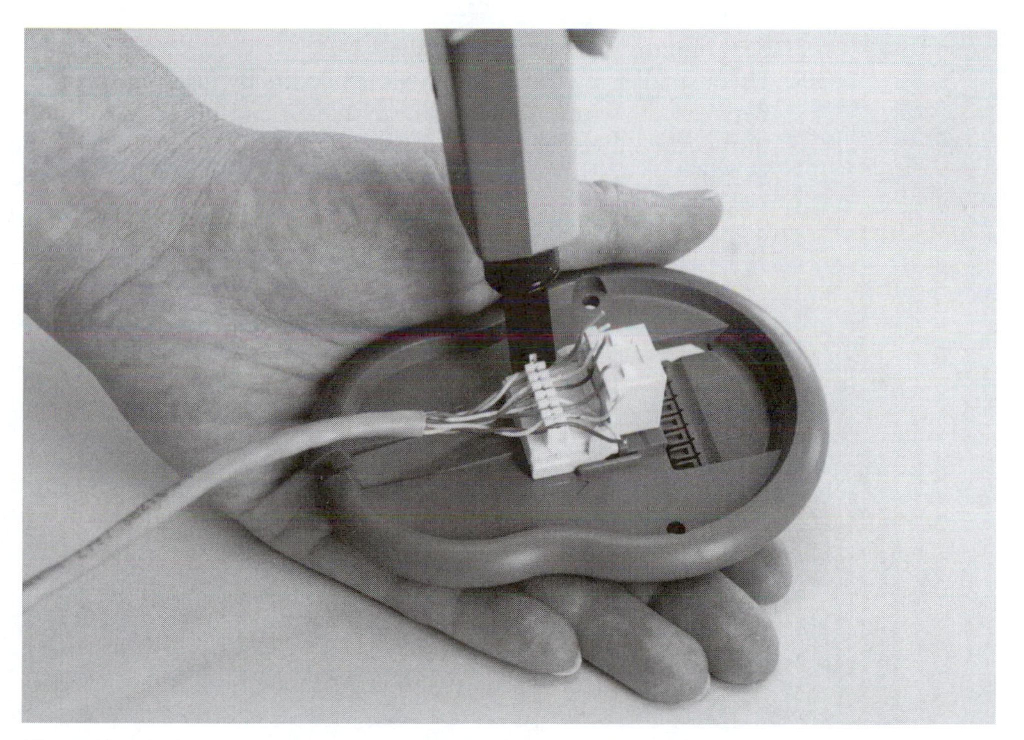

**Figure 9–19.** Special holders make punching down jacks much easier.

### Terminating RJ-45 8-pin Modular Plugs

The final hands-on exercise is to make two patch cables to complete the 568 "channel." In most networks, you buy patchcords made in a factory, and they are made with stranded wire, not solid wire like regular Cat 5 cable. But in a pinch, you can easily make your own patchcords. We will use the cable provided and modular connectors, usually called plugs, to make two of our own patchcords.

Most 8-pin modular plugs use the same termination process. You will strip the cable jacket, untwist the wire pairs, align the wires according to the proper color code for the termination style (568A, 568B, USOC, and so on), cut to 1/2-inch length, inset in the connector, and crimp to set the IDC connectors. *As with jacks and punchdowns, it is mandatory to maintain the twists as closely as possible to the termination!*

#### *Exercise*

Tools: Jacket stripper, cable cutter, modular connector crimper
Materials: 3 meters of Cat 5 cable, modular connectors

1. Use the jacket stripper to strip approximately 2 inches of jacket off the cable.
2. Separate the pairs by fanning them out.
3. Use a small screwdriver to untwist the pairs by inserting the screwdriver between the wires in the pair near the cut end of the jacket and pulling outward, untwisting the wires back to the end of the jacket (Figure 9–20).
4. Use your fingers to straighten the wires as much as possible (Figure 9–21).

**Figure 9–20.** For plugs, untwist the pairs all the way back to the jacket.

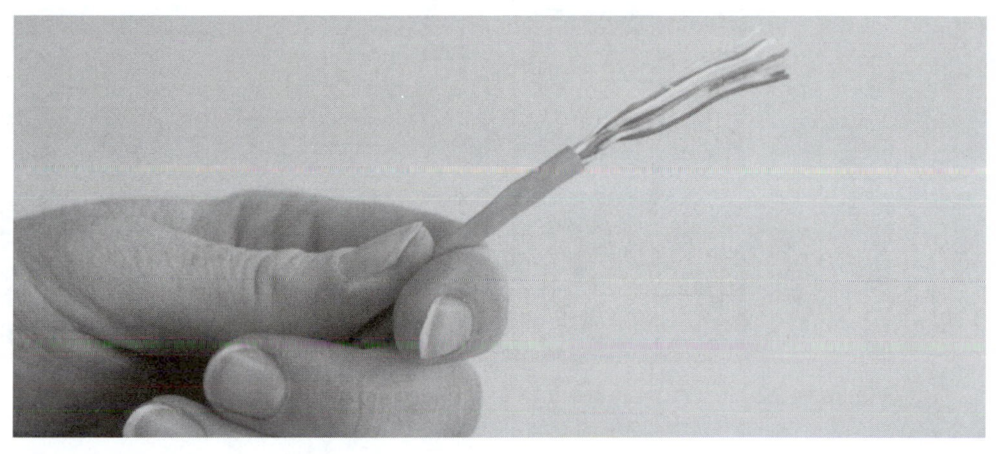

**Figure 9–21.** Straighten the wires and place in the proper order.

5. Place the wires in color-coded order for a T568A or T568B connector, as shown in Table 9–1. *Note:* You can use either the T568A or T568B pair configuration, as long as you use the same configuration for both ends of the patchcord.

**Table 9–1** Color Code Order for T568A and T568B Plugs

| Pin Number | T568A Color Codes | T568B Color Codes |
|---|---|---|
| 1 | W/G | W/O |
| 2 | G | O |
| 3 | W/O | W/G |
| 4 | BL | BL |
| 5 | W/BL | W/BL |
| 6 | O | G |
| 7 | W/BR | W/BR |
| 8 | BR | BR |

6. When you have aligned the wires correctly, hold them flat between your fingers and wiggle them back and forth to make them as flat and straight as possible.
7. Using the cable cutter, cut the wires to 1/2-inch length. Make sure the wires do not get out of order!
8. Place the connector onto the cable by inserting the wires into the connector channels and sliding them all the way in (Figure 9–22). Make sure the wires butt up against the inside front wall of the connector.

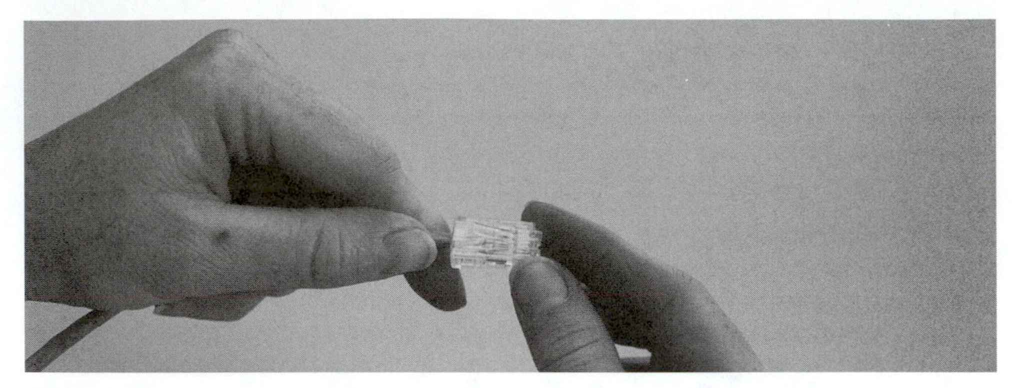

**Figure 9–22.** Insert the wires into the connector fully.

9. Place the assembled connector into the crimper with jaws for the RJ-45. When the connector is properly positioned, squeeze the tool handles to allow one full ratchet cycle until the tool completely closes and opens again (Figure 9–23).
10. To remove the connector after crimping, press down on the connector key and pull the connector out of the tool. Finally, inspect the completed crimp. All pins should be fully crimped and of the same height.

Repeat the process on the other end of the cable to produce a patchcord. Then make several more patchcords to use in testing your cabling installation.

### Terminating Cat 3 Cable for Telephones

The Cat 3 installation for telephones is almost exactly like the Cat 5 for LANs, except you will usually use a "type 66" block (Figure 9–24). The 66 block has rows of four punchdown positions, with each side being a permanently connected pair. In

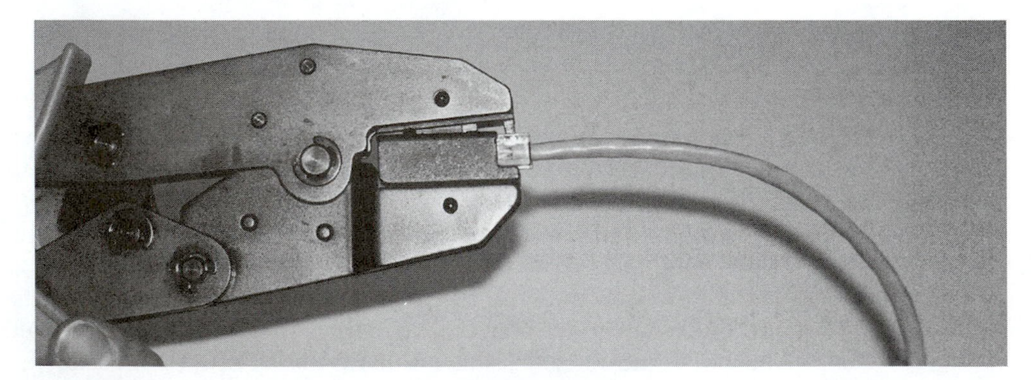

**Figure 9–23.** Crimp the plug fully with the crimper.

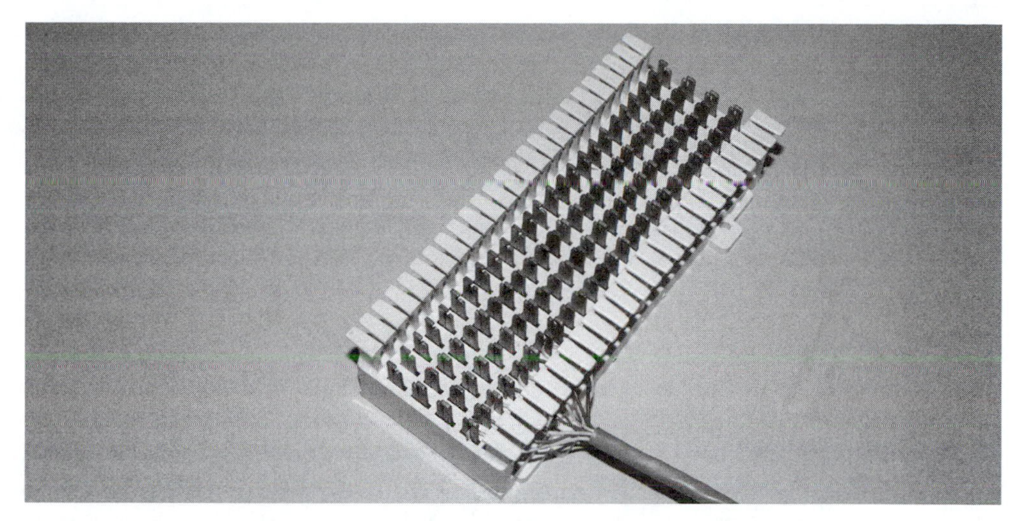

**Figure 9–24.** Type 66 punchdown block.

use, the outside positions are used to terminate cable pairs, while the inside positions are used to terminate cross-connect wires. If the connection is directly across the row, a bridging clip can be used to save wiring time (Figure 9–25).

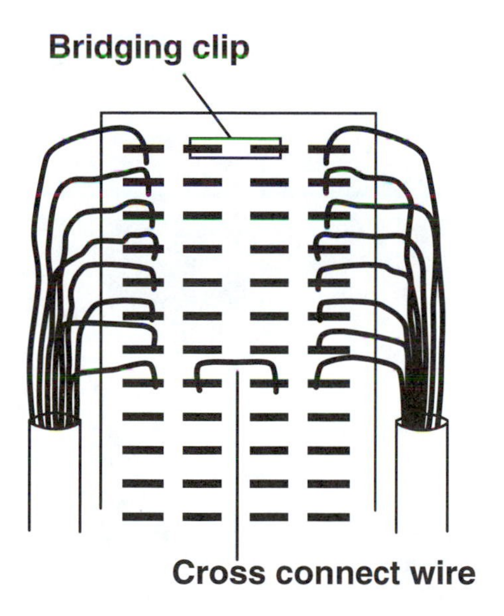


**Figure 9–25.** 66 block termination for telephones.

Since we are terminating for telephone connections over Cat 3, not LAN, cabling, we can remove more of the cable jacket and allow more untwisted wire without harming the performance of the cabling. Remember, however, that this is the case only for cabling that is used exclusively for telephone service, never for cabling that is for LAN usage!

The tools required terminating the 66 block are a jacket stripper and a punch-down with 66 blade (*cut* side out). The procedure is as follows:

1. Identify the Cat 3 cable in the training box.
2. Strip off about 4 inches of jacket.
3. Untwist the pairs, but keep pairs together.
4. Following the same color code as Cat 5 (remember BLOG? blue-orange-green, then brown, white wire first), punch down each wire into a position on the end of one row on the block (Figure 9–26).

The punchdown tool works almost exactly like it did on the 110 block. Make sure you use the 66 blade end that says "cut," so it will cut off the excess wire. And make sure the "cut" side is on the side of the position that has the excess wire to be cut off!

After you have terminated one Cat 3 cable to the 66 block, terminate the other Cat 3 cable on the other side of the block with the color-coded wires in the same rows.

*Cross-connecting.* Now you can use bridging clips to connect the rows to complete the connection. As another option, if you want more practice doing punchdowns, you can cut a short length of cable, pull out the wires, untwist them, and then use them to punch down on the inner positions as cross-connect wires.

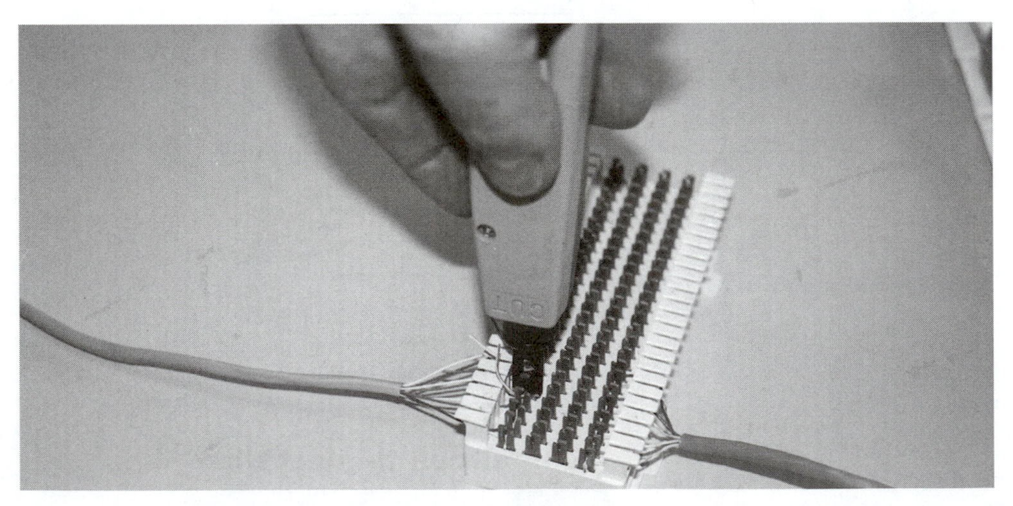

**Figure 9–26.** Punchdown cables to be connected across from each other if possible.

*Terminating the outlet jacks.* Following exactly the same procedure as used for the Cat 5 jacks, terminate the Cat 3 cable into the jacks provided and snap them into the outlet plates in the telephone position. Remember, you will need to change blades on the punchdown tool to a 110 blade to match the terminations on the jacks.

*Descriptions of other terminating types.* Some manufacturers have developed terminations for jacks and patch panels that do not require punchdowns. These have built-in IDC terminations that merely require stripping the jacket, cutting the wire to length, and inserting the wire in the proper receptacle. A simple push on a block or crimp with pliers sets the IDC connector, with no other tools required.

Each of these blocks is unique, but similar. If you use them, read the manufacturers' directions carefully and follow them exactly to ensure proper termination.

*Testing and troubleshooting the cabling network.* Now that you have completed terminating the cables, you must test them. The first stage of testing is to use an instrument called a wire mapper to test for wire map. If you are testing for full Cat 5 performance, you will need a Cat 5 tester also. Most installers use wire mappers for preliminary testing, often providing one to each installer, so wire map problems can be corrected before the expensive Cat 5 tester is used for performance testing. Since 3 to 5 percent of the links typically have some wire map problem, it is usually more cost-effective to find wire map problems before using the Cat 5 tester.

**Tools: The Wire Mapper**

The wire mapper tests for all the wire mapping faults found in Cat 3 or Cat 5 cabling, including opens and shorts, reversed pairs, crossed pairs, and split pairs. Instrument display indicates the problem and the pair(s) at fault. The tester consists of the instrument itself, a remote terminator, and several short Cat 5 cables and dual jack terminators to allow connections to either plugs or jacks. For field use, coded terminators are available with numbers that allow identification of individual cables for testing convenience.

*Testing patchcords.* To use the wire mapper to test your patchcords, connect the wire mapper directly to the terminator with one of your patchcords and a dual jack or female-female modular 8-pin connector (Figure 9–27). Initiate the test sequence on the wire mapper. When finished testing, the wire mapper will show the results for a few seconds and then automatically turn off. If it does not indicate that all pairs are correctly wired, check another patchcord or terminator, since valid testing will require properly wired patchcords and terminators.

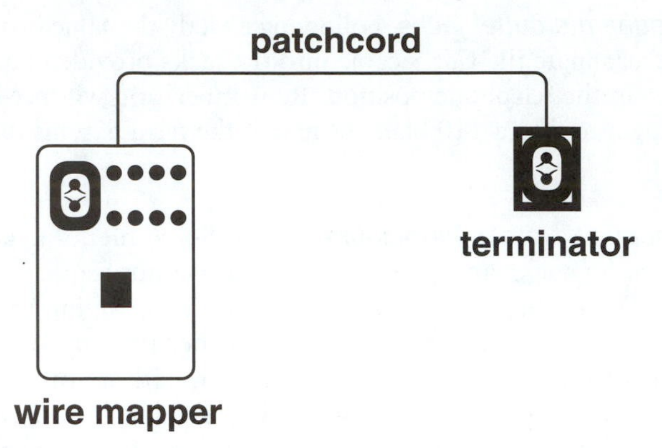


**Figure 9–27.** Checking patchcords with the wiremapper.

*Training box link testing.* To test your simulated cabling network in the "channel" configuration, connect the wire mapper and the terminator to each end of the cables by using the test patchcords to connect into the jacks at either end of the training board. Push the test button and test for correct wire connections. You can test the "link" portion of the installed cable by plugging the terminator directly into the jack at one end of the simulated network (Figure 9–28).

### Troubleshooting Problems

Suppose the wire mapper indicates that you have a problem with a cable or a link. The following text discusses what the errors mean, using diagrams of the problems applied to a 568A connection.

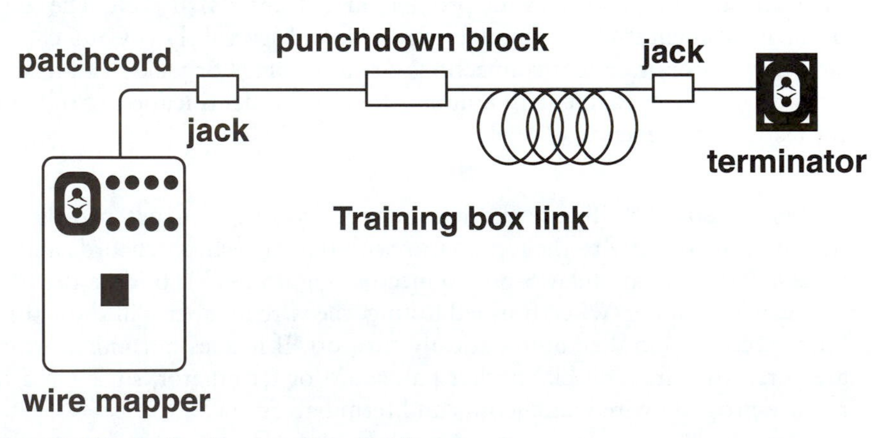


**Figure 9–28.** Testing the complete Cat 5 channel.

*Open or shorted pairs.* The wire mapper will indicate a short where two conductors are accidentally connected or an open where one or more wires are not connected to the pins on the plug or jack. Opens can also occur due to cable damage. Most shorts and opens will occur at the connections. Physical examination of the connections should find the fault. If possible, unpatch the cables in a link to help find the location. A time domain reflectometer (TDR) test will show the distance to the fault to assist you in locating it.

In the example in Figure 9–29, the wire mapper will show a "short" for pair 3 (1/2) and an "open" for pair 2 (3/6). Pair 1 (4/5) will show an "open-crossed," indicating that it is open but one wire is crossed, while pair 4 (7/8) will show "crossed," indicating it is crossed with another pair.

Shorts can also occur across two pairs. In the example in Figure 9–30, the wire mapper will show "crossed" for both pair 4/5 and pair 7/8.

*Reversed pairs.* Reversed pairs occur when the conductors (Tip and Ring) are reversed on one end of the pair. In the example in Figure 9–31, the wire mapper will indicate "reversed" on the 3/6 pair (pair 2 ). Reversal can occur at any connection, so look for the correct wire color coding at every termination point. Since it involves reversing white and a color, it should be easy to find.

*Crossed pairs.* Transposed or crossed pairs occur where both conductors of one pair are reversed with both conductors of another pair at one end. In the example in Figure 9–32, the wire mapper will indicate "crossed" on both the 1/2 and 3/6 pairs. If you see a crossed pair error involving pairs 2 and 3, it may be that the link has a mixture of 568A and 568B terminations, as these two terminations differ by the reversal of these two pairs.

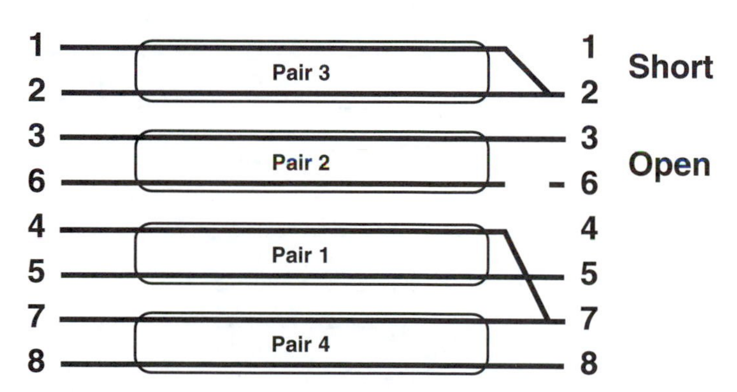


**Figure 9–29.** Opens or shorts can be at either end of the cable.

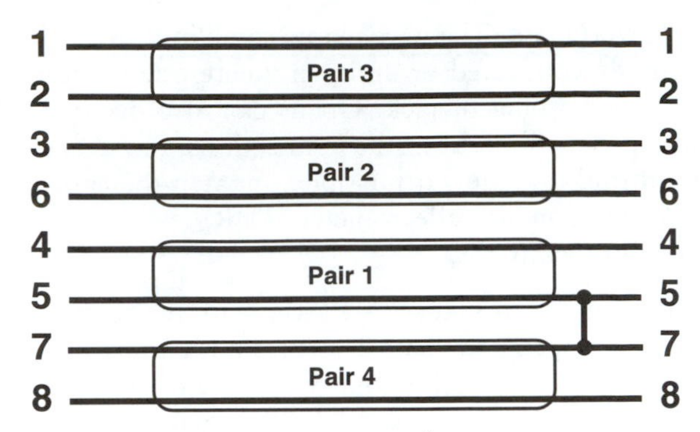


**Figure 9–30.** Shorts can also occur across two pairs.

*Split Pairs.* Some wiring verifiers will provide basic connection information, but some faults like split pairs may not show up in their wire maps. Split pairs occur where one wire each of two pairs are reversed on *both* ends. This is impossible to find with a normal wiring verifier, since the wire map is correct—that is, the pin connections are correct, but the wires are not in proper pairs. This error can only be detected in a crosstalk (NEXT) test or balance test, where the unbalanced pairs can be detected. In the example in Figure 9–33, the wire mapper will indicate "split" on both 1/2 and 3/6 pairs.

*Troubleshooting wire map problems.* Some errors are easier to trace than others. Random problems require simply rechecking all terminations in sequence. If you have a Cat 5 tester, special cable testers, or a TDR (time domain reflectometer), you can ascertain the distance to opens and shorts. If you see a crossed pair error involv-

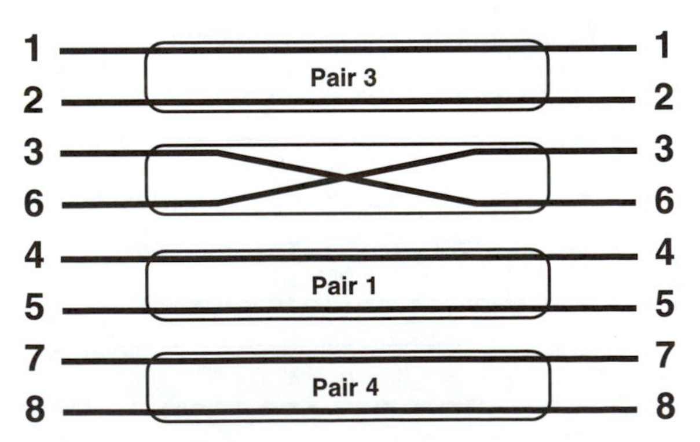


**Figure 9–31.** Reverse wires on pair 2.

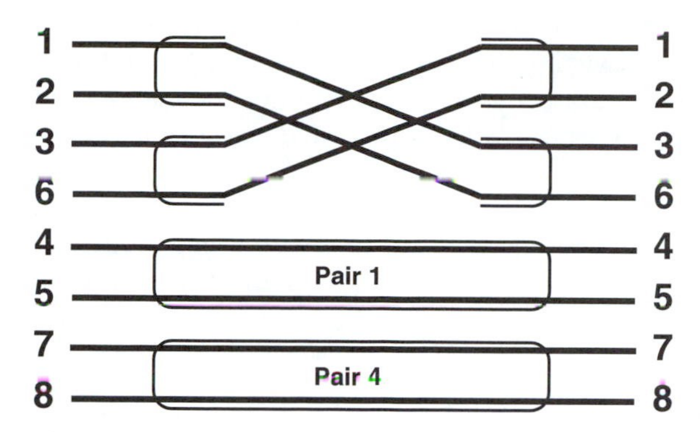


**Figure 9–32.** The wire mapper will indicate "crossed" on both the 1/2 and 3/6 pairs.

ing pairs 2 and 3, it may be that the link has a mixture of T568A and T568B terminations, as these two terminations differ by the reversal of these two pairs. If pair 1 is *reversed* and everything else is in error, a USOC pinout might be involved somewhere.

### Category 5 Testing

If a Category 5 tester is available, test your cable link with it. The Cat 5 tester does much more than a wire map; it measures length, attenuation, crosstalk (NEXT, or near end crosstalk), and ACR (attenuation crosstalk ratio) per EIA standards. Refer to the chapter on wire testing for a complete explanation of all these tests. Generally, Cat 5 testers are set up to provide a "pass/fail" test for the link and only display more information if the link fails a test.

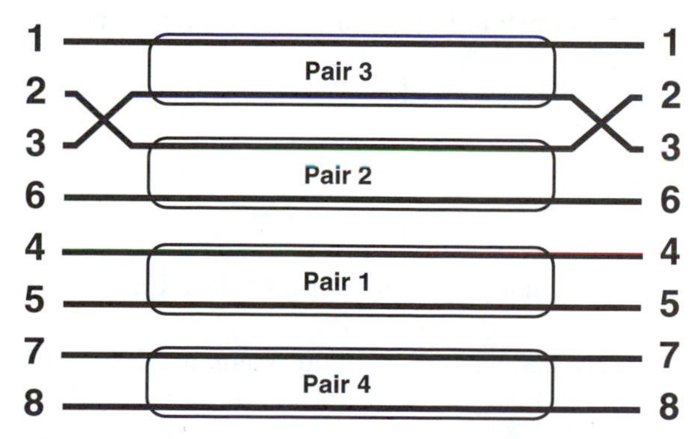


**Figure 9–33.** The wire mapper will indicate "split" on both the 1/2 and 3/6 pairs.

The most common failure in Cat 5 performance testing is NEXT, generally caused by untwisting the pairs too much at a punchdown block or jack. Most other performance parameters are a function of the manufacture of the cable, so no failures should be expected from pretested Cat 5 cable.

If you have a NEXT failure, check pairs at all punchdowns, including the back of jacks, for too much untwisted wire. If that does not appear to be the problem, look for bends or kinks in pairs. If you are using modular adapters to connect patchcords, make certain they are Cat 5 rated, as normal phone connectors will always fail NEXT.

## CHAPTER REVIEW

1. For coax CATV cable, what is the most important part of installation?
2. Name all the parts of the coaxial cable construction.
3. What tools do you use to terminate coaxial cable?
4. Since the coaxial cable is well-protected and durable, one can pull as hard as one wishes to install the cable. T or F
5. What will poor termination of coax cable cause?
6. Can Cat 5 cable be used in place of Cat 3? Yes or No
7. Can Cat 3 cable be used in place of Cat 5? Yes or No
8. What is the major installation problem with Cat 5 cable?
9. What are the four colors of the wires in Cat 5 cable?
10. What is the maximum pulling tension of Cat 5 cable?
    A. Your weight
    B. The weight of the new box of cable
    C. 25 pounds
    D. The weight of your car
11. What is the longest horizontal distance Cat 5 cable should be from the patch panel to the work area outlet? (meters and feet)
12. What tools do you use to punch down the wires into a 110 block?
13. How far should you untwist the pairs in Cat 5 cable to terminate the ends?
14. In a 110 block, what makes the contacts between the two wires in the block?
15. What two ways can you use to mate cables in a 110 block?
16. What style of punchdown block is commonly used with Cat 3 cable for telephones?
17. A wire mapper is not used to test Cat 5 compliance up to 100 MHz. T or F
18. Where do most shorts and opens occur?

# PART 3

# FIBER OPTICS

CHAPTER

# 10

# FIBER OPTIC CABLING SYSTEMS

## WHICH FIBER OPTICS?

Fiber is used in almost every telephone system, most CATV systems, many LANs, some security (CCTV) systems, and many more applications. Each of these systems has unique designs and requirements. We can divide fiber optics into two different applications, outside plant and premises applications, which are very different.

Telco and CATV systems are called "outside plant" systems, since much of the cabling is installed in the outdoors, either underground or aerial. Virtually all outside plant networks use single-mode fiber and laser transmitters operating at high speeds. Cables are spliced into long continuous runs, mostly by fusion splicing, and must be tested with OTDRS (optical time domain reflectometers) to ensure proper installation. Outside plant installations involve choosing the proper cable to withstand the environment and careful pulling of the cable to prevent damage during installation. Outside plant installations are also expensive, involving expensive equipment, travel, and lots of time to install cable properly.

Premises installations of fiber optics are much simpler. Applications like LANs or security use multimode fiber and LED transmitters in short runs, usually indoors in a benign environment that allows use of simple, low-cost cables. Splices are rare and terminations are usually done on-site. Hardware and tools are inexpensive, and skills are easily mastered, making it possible for most installers to do fiber alongside copper cabling.

Some applications use a little of both premises and outside plant technology. Utilities run fiber around cities and over long distances for communications and grid management. Cities have adopted fiber for smart traffic signal controls. College campuses have lots of short links for backbones but often use longer links between buildings. Military and government installations are similar to campuses but on a larger scale.

## Fiber or Copper?

Every manufacturer of networking or data communications equipment offers fiber optic interfaces as well as those operating on UTP copper cable. Fiber is necessary where longer distances or electromagnetic interference (EMI) is a problem. It is also preferred by network users who are planning on upgrades to higher bandwidth in the future. Today, fiber is used in the majority of LAN backbones and in all telephone and CATV networks.

Part of the popularity of fiber comes from its performance characteristics. High bandwidth, distance capability, and noise immunity make it possible to use fiber where copper is not possible or cost-effective. Another factor in fiber popularity is its ease of installation and testing.

Almost twenty years ago, the PhD's from Bell Labs installed all the fiber optics. Today, fiber is so easy to install that electrical contractors install much of it, because fiber cable, connector, and splice manufacturers have been working furiously over the last few years to make their components easier to install and less expensive.

Installing fiber is easy with some basic training and practice, and testing is much easier and less expensive than copper wire. Even though it is glass fiber, it has more strength and greater tolerance to abuse than Cat 5 copper wire. And plastic fiber may soon offer more performance than Cat 5 at equal or lesser prices.

## Fiber optic Standards

Watch out for confusing standards! The only universal, mandatory standard for fiber optic cable plant installation is Chapter 770 of the *National Electrical Code*® (*NEC*®). There are many of references to EIA/TIA 568, Bellcore, and a number of other "standards," but these are voluntary. They are the result of manufacturers getting together to decide how they want their products used, with little input from installers or end users.

For many years, EIA/TIA has been developing standards for fiber optic components and appropriate test procedures that are widely recognized by manufacturers. Adherence to these standards means that most component specifications are comparable and components like fiber and connectors can be intermated. Other groups like The Institute of Electrical and Electronic Engineers (IEEE) and the American National Standards Institute (ANSI) have developed network standards that ensure interoperability among network products from vendors who follow these guidelines.

ANSI has chartered the Fiber Optic Association (FOA) and the National Electrical Contractors Association to develop installation standards for fiber optics, covering workmanship standards for installations. Remember that all these standards are guidelines—not regulations. If someone asks you about standards, remind them they can specify anything they want, but only *NEC*® 770 is mandatory. Otherwise common sense should rule.

## Getting Started

It is not hard to get started with fiber optic installations, nor is it expensive, especially if you already know how to install copper wire. You need some special tools and some test gear (much less expensive than copper testers). Many of the tools used in working with fiber optic cables are the same ones used with copper cables.

Most beginners need some training. Although it is possible to learn by buying some equipment and parts and experimenting on your own, it is faster and cheaper in the long run to get some training. You do not necessarily have to go to a class, since with self-study courses, you can learn at home on your own time.

But you must maintain your skills by practicing continually and learning how to use new connectors, new splices, and even new test gear that is produced. Take advantage of free training offered by manufacturers and distributors whenever you can. Fiber optics is still a fast-moving technology, so you must work at keeping up-to-date.

## Safety

When one mentions safety in fiber optic installation, the first image that comes to most people's minds is a laser burning holes in metal—or in your eyeball. This image has little relevance to fiber optic communications systems. Optical sources used in fiber optics generally are of much lower power levels and are not focused into a small spot like other applications.

In fact, most data communications links use LEDs of very low power levels, and even the lasers in most fiber optic installations are of relatively low power. The light that exits an optical fiber is also spreading out in a cone, so the farther away from the end of the fiber you are, the lower the amount of power striking a given-sized spot. Eye safety is only a concern in high-powered CATV systems.

Fiber optic installation, however, is not without risks. You will be continually exposed to small scraps of bare fiber cleaved off the ends of the fibers being terminated or spliced. These scraps are very dangerous. The cleaved ends are extremely sharp and can easily penetrate your skin. If they get into your eyes, they are very hard to flush out. Always wear safety glasses when working with bare fibers.

Avoid painful accidents by exercising a little caution. Dispose of all scraps properly. Keep a piece of double-stick tape on the bench to stick scraps to, or put them in a properly marked paper cup or other container to dispose of later. Do not drop them on the floor where they will stick in carpets or shoes and be carried elsewhere. Do not eat or drink anywhere near the work area.

Fiber optic splicing and termination use various chemical cleaners and adhesives as part of the processes. Normal handling procedures for these substances should be observed. Even simple isopropyl alcohol, used as a cleaner, is flammable and should be handled carefully.

## FIBER OPTIC NETWORKS

Fiber optic transmission systems all work in a manner similar to the link diagram shown in Figure 10–1. They consist of a transmitter that takes an electrical input and converts it to an optical output from a laser diode or LED and couples the light into an optical fiber. The light from the transmitter is coupled into the fiber with a connector and is transmitted through the fiber optic cable plant. The light is ultimately coupled to a receiver where a detector converts the light into an electrical signal that is then conditioned properly for use by the receiving equipment.

Each duplex link consists of two separate links operating on two fibers transmitting in opposite directions. Methods exist to operate bidirectionally on a single fiber, but it is less cost-effective than two separate fiber links.

Just as with copper wire or radio transmission, the performance of the fiber optic data link can be determined by how well the reconverted electrical signal out of the receiver matches the input to the transmitter.

The ability of any fiber optic system to transmit data ultimately depends on the optical power at the receiver as shown in Figure 10–2, which shows the data link bit

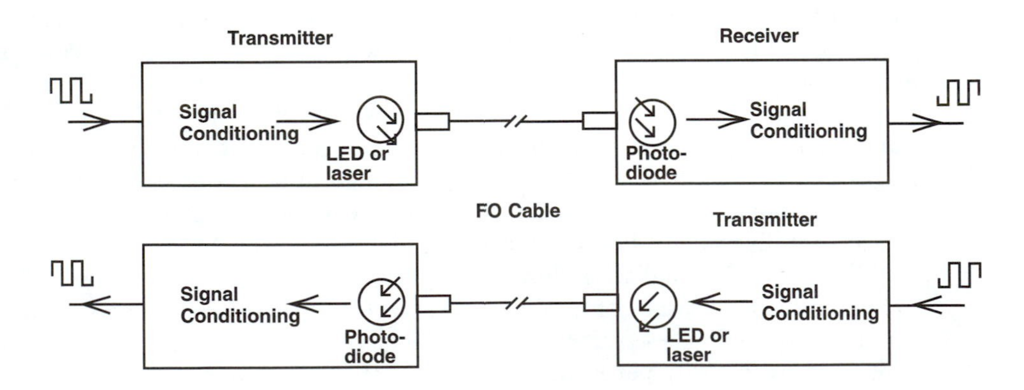


**Figure 10–1.** A typical duplex fiber optic link.

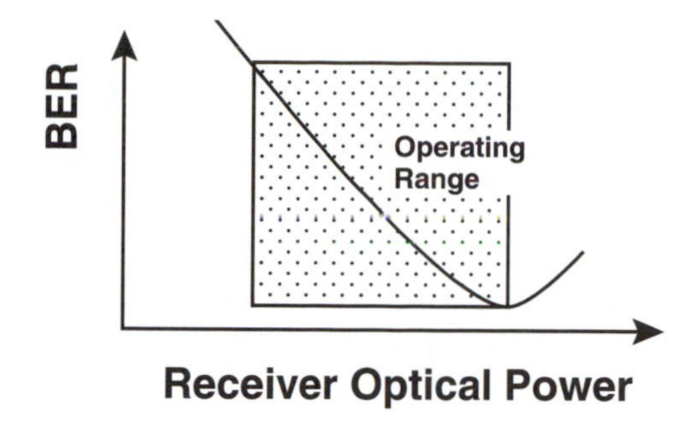


**Figure 10–2.** The performance of a fiber optic link depends on the power level at the receiver.

error rate as a function of optical power at the receiver. Either too little or too much power will cause high bit error rates. Too much power, and the receiver amplifier saturates and distorts the signal; too little, and noise becomes a problem. This receiver power depends on two basic factors: how much power is launched into the fiber by the transmitter and how much is lost by attenuation in the optical fiber cable that connects the transmitter and receiver.

Data links can be either analog or digital in nature. Both have some common critical parameters and some major differences. For both, the optical loss margin is most important. This is determined by connecting the link with an adjustable attenuator in the cable plant and varying the loss until one can generate the curve shown in Figure 10–2. Analog data links will be tested for signal-to-noise ratio to determine link margin, while digital links use bit error rate as a measure of performance. Both links require testing over the full bandwidth specified for operation, but most data links are specified for a particular network application—like AM CATV or CCTV for analog links, and SONET, Ethernet, FDDI, or ESCON for digital links.

The optical power margin of the link is determined by two factors, the sensitivity of the receiver, which is determined in the bit error rate curve, and the output power of the transmitter into the fiber. The minimum power level that produces an acceptable bit error rate determines the sensitivity the receiver. The power from the transmitter coupled into the optical fiber determines the transmitted power. The difference between these two power levels determines the loss margin of the link.

If the link is designed to operate at differing bit rates, it is necessary to generate the performance curve for each bit rate. Since the total power in the signal is a function of pulse width, and pulse width will vary with bit rate (higher bit rates mean shorter pulses), the receiver sensitivity will degrade at higher bit rates.

Every manufacturer of data links components and systems specifies them for receiver sensitivity (perhaps a minimum power required) and minimum power coupled into the fiber from the source. Typical values for these parameters are shown in Table 10–1. In order to test them properly, it is necessary to know the test conditions. For data link components, it includes input frequency or bit rate and duty cycle, power supply voltages, and the type of fiber coupled to the source. For systems, it will be the diagnostic software needed by the system.

Most applications of fiber optics can be summarized in the categories presented in Table 10–1. Although all telecom and CATV systems are singlemode, applications for singlemode (SM) datacom networks are becoming more common, especially in metropolitan area networks (MANs), campuses, traffic control systems, and utility power management systems that cover large geographic areas. These SM datacom systems use telecom-type links with specifications similar to those shown for telecom systems.

**Table 10–1** Typical Fiber Optic Link/System Performance Parameters

| Link type | Source/Fiber Type | Wavelength (nm) | Transmit Power (dBm) | Receiver Sensitivity (dBm) | Margin (dB) |
|---|---|---|---|---|---|
| Telecom | laser/SM | 1300 | +3 to -6 | -40 to -45 | 34 to 48 |
| | | 1550 | 0 to -10 | -40 to -45 | 40 to 45 |
| Datacom | LED/MM | 850 | -10 to -20 | -30 to -35 | 10 to 25 |
| | | 1300 | -10 to -20 | -30 to -35 | 10 to 25 |
| CATV (AM) | laser/SM | 1300 | +10 to 0 | 0 to -10 | 10 to 20 |

Telecom systems normally use 1,300-nm links for shorter runs and 1,550 nm for longer distances, like intercity and submarine cables. The 1,550-nm band is also popular for wavelength-division multiplexing, where several lasers of different wavelengths share a single fiber, and fiber amplifiers, which are all-optical repeaters. CATV has also experimented with 1,550-nm transmission, to be able to use fiber amplifiers to boost signal levels, then split the signal to many different nodes.

Within the data communications links and networks, there are many vendor-specific fiber optic systems, but there are also a number of industry standard networks. These networks have agreed-upon specifications common to all manufacturers' products to ensure interoperability. Table 10–2 shows a summary of these systems.

The newest network specified for fiber optics is Gigabit Ethernet (Table 10-3). It has been well received as a new high-speed migration path for backbones because of its familiar architecture and protocol. Gigabit Ethernet will probably be able to use UTP cabling also, but over a very limited length. Three fiber solutions are included in the draft standard, all using lasers for their higher bandwidth capability. The tentative specifications for Gigabit Ethernet are detailed in Table 10–3, including the short wavelength link using a new type of inexpensive laser, a VCSEL, (vertical cavity surface emitting laser).

Like other high-speed links operating over multimode fiber, Gigabit Ethernet will be limited in distance by the bandwidth of the fiber, not the loss budget. New versions of 62.5/125 fiber and 50/125 micron fiber with higher bandwidth will offer distance capability much greater than the standard distances.

**Table 10–2** Fiber-optic Datacommunications Networks

| Network | IEEE802.3 FOIRL | IEEE802.3 10base F | IEEE802.5 Token Ring | IEEE802.12 100BaseF | ANSI X3T9.5 FDDI | ESCON IBM |
|---|---|---|---|---|---|---|
| Bit rate (MB/s) | 10 | 10 | 4/16 | 100 | 100 | 200 |
| Architecture | Link | Star | Ring | Star | Ring | Branch |
| Fiber type | MM, 62.5 | MM, 62.5 | MM, 62.5 | MM, 62.5 | MM/SM | MM/SM |
| Link length (km) | 2 | 2 | — | 2 | 2/60 | 3/20 |
| Wavelength (nm) | 850 | 850 | 850 | 1300 | 1300 | 1300 |
| Margin (dB, MM/SM) | 8 | 12.5 | 12 | 11 | 11/27 | 8*(11)/16 |
| Fiber BW (MHz-km) | 150 | 150 | 150 | 500 | 500 | 500 |
| Connector | SMA | ST | FDDI | ST | FDDI | ESCON |

*IBM specifies a nonstandard method of testing cable plant loss that reduces the loss to a maximum of 8 dB. However, the component specifications are similar to FDDI, so testing to FDDI margins is appropriate.

**Table 10–3** Gigabit Ethernet Specifications (Tentative)

| Network | IEEE802.3z 1000BaseSX | IEEE802.3z 1000BaseLX | IEEE802.3z 1000BaseLX |
|---|---|---|---|
| Bit rate (MB/s) | 1000 | 1000 | 1000 |
| Architecture | Star | Star | Star |
| Fiber type | MM, 50/62.5 | MM, 50/62.5 | SM |
| Link length (km) | 0.22 to 0.5 | 0.55 | 5 |
| Wavelength (nm) | 850 VCSEL | 1300 Laser | 1300 Laser |
| Margin (dB, MM/SM) | TBD | TBD | TBD |
| Fiber BW (MHz-km) | 160/400 | 500 | NA |
| Connector | SC | SC | SC |

## CHAPTER REVIEW

1. When is it necessary to use fiber optics in communications systems?
2. What connector do most users install in fiber optics?
3. What connector is the EIA/TIA standard?
4. What is another name for "zone cabling"?
5. What can make fiber optics competitive to copper?
6. What is the only *mandatory* standard for fiber optic installations?
7. Who chartered the Fiber Optic Association (FOA) and the National Electrical Contractors Association to develop standards?
8. Are standards the same as regulations or codes?
9. Which test equipment costs less: copper or fiber?
10. For safety in fiber optics, what part of the body should be of most concern? Why?

11. What determines how well the fiber optic system works?
12. Why are two fibers used for duplex transmission?
13. What two factors does the receiver power depend on?
14. Which of the following applications use analog links and which use digital links?

| Application | Analog | Digital |
|---|---|---|
| AM CATV | | |
| CCTV | | |
| SONET | | |
| Ethernet | | |
| FDDI | | |
| ESCON | | |

15. What is the difference between transmitter power output and receiver sensitivity called?
16. What happens to the receiver at high power levels?
17. What determines the sensitivity of the receiver?
18. What happens to the receiver sensitivity at higher bit rates?
19. What sources and wavelengths are used for longer distances in telecom?
20. What new device is used for 850-nm Gigabit Ethernet?

CHAPTER

# 11

# OPTICAL FIBER AND CABLE

Optical fiber is composed of a light carrying core and a cladding that traps the light in the core, causing total internal reflection (Figure 11–1). Most fiber is composed of a solid glass core and cladding, with a plastic buffer coating for protection from physical damage and moisture. The plastic buffer is stripped from glass fiber for terminating or splicing. Some fibers have a glass core and plastic cladding, and some are all plastic.

Fiber has two basic types, multimode and singlemode (Figure 11–2). Multimode fiber means that light can travel many different paths (called modes) through the core of the fiber, which enter and leave the fiber at various angles. The highest angle that light is accepted into the core of the fiber defines the numerical aperture (NA).

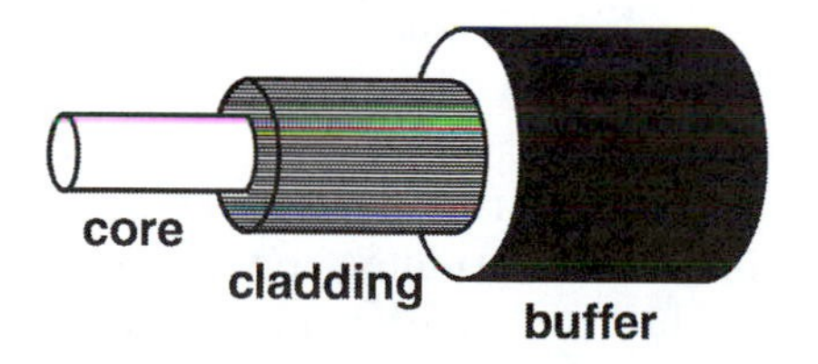


**Figure 11–1.** Optical fiber consists of a core, cladding, and a protective buffer coating.

**Multimode Step Index**

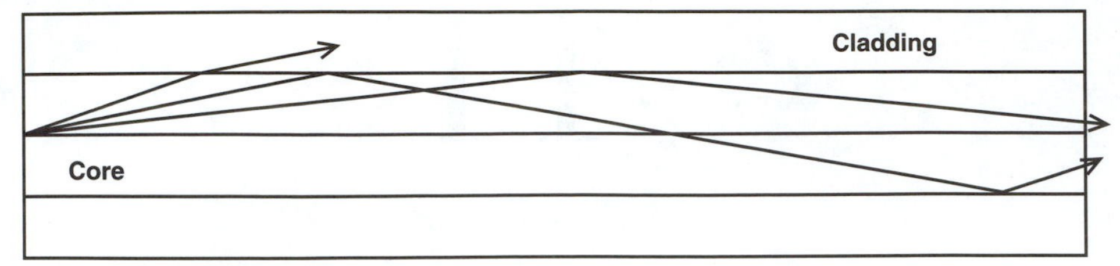


**Multimode Graded Index**

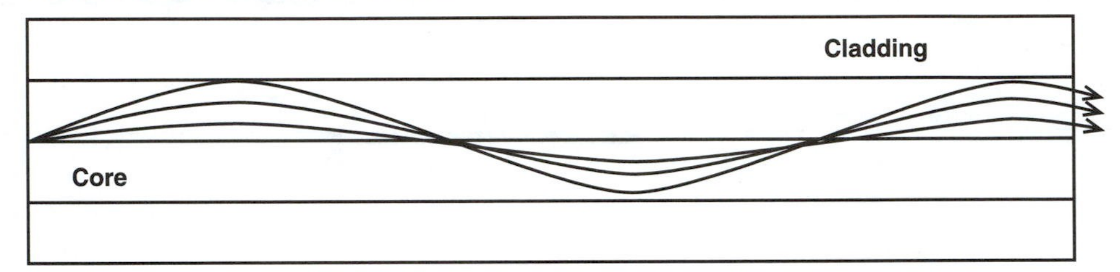


**Singlemode**

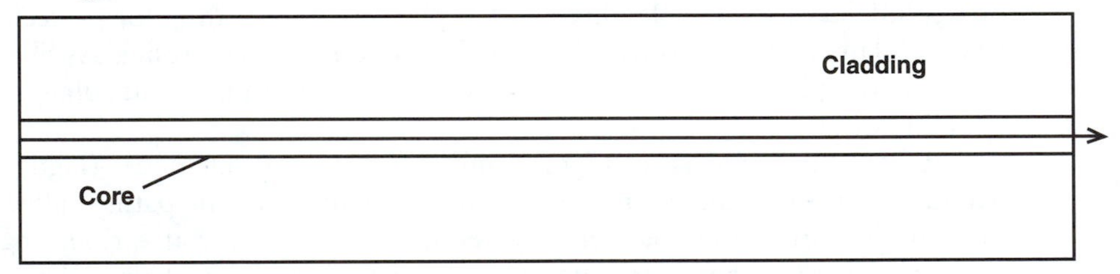


**Figure 11–2.** The three types of optical fiber.

Two types of multimode fiber exist, distinguished by the index profile of their cores and how light travels in them. Step index multimode fiber has a core composed of one type of glass. Light traveling in the fiber travels in straight lines, reflecting off the core/cladding interface. The NA is determined by the differences in the indices of refraction of the core and cladding and can be calculated by Snell's law. Since each mode or angle of light travels a different path link, a pulse of light is dispersed while traveling through the fiber, limiting the bandwidth of step index fiber.

In graded index multimode fiber, the core is composed of many different layers of glass, chosen with indices of refraction to produce an index profile approximating a parabola. Since the light travels faster in lower index of refraction glass, the light will travel faster as it approaches the outside of the core. Likewise, the light traveling closest to the core center will travel the slowest. A properly constructed index profile will compensate for the different path lengths of each mode, increasing the bandwidth capacity of the fiber by as much as one hundred times that of step index fiber.

Singlemode fiber just shrinks the core size to a dimension about six times the wavelength of the fiber, causing all the light to travel in only one mode. Thus modal dispersion disappears, and the bandwidth of the fiber increases by at least another factor of 100 over graded index fiber.

Each type of fiber has its specific application, and its performance characteristics are tailored to that application (Table 11–1). Step index fiber is used where large core size and efficient coupling of source power is more important than low loss and high bandwidth. It is commonly used in short, low-speed data links with LED sources. It may also be used in applications where radiation is a concern, since it can be made with a pure silica core that is not readily affected by radiation.

**Table 11–1** Fiber Types and Typical Specifications

| Fiber Type | Core/Cladding Diameter (μm) | Attenuation Coefficient (dBkm) | | | Bandwidth (MHz-km) |
|---|---|---|---|---|---|
| | | 850 nm | 1300 nm | 1550 nm | |
| Step index | 200/240 | 6 | | NA | 50 @ 850 nm |
| Multimode | 50/125 | 3 | 1 | NA | 600 @ 1300 nm |
| Graded Index | 62.5/125 | 3 | 1 | NA | 500 @ 1300 nm |
| | 85/125 | 3 | 1 | NA | 500 @ 1300 nm |
| | 100/140 | 3 | 1 | NA | 300 @ 1300 nm |
| Singlemode | 9/125 | | 0.5 | 0.3 | High |
| Plastic (POF) | 1 mm | (0.2 dB/m @ 665 nm) | | | Low |

Most multimode links use 850- or 1,300-nm LEDs, since the larger core readily accepts the broad output pattern of the LED. LEDs are limited to speeds of up to 200 MB/s, so they are not appropriate for very high-speed links. Few multimode applications use lasers, since the coherent light output of a laser may have problems with the many modes propagated in the fiber. A new type of laser, called a VCSEL (vertical cavity surface emitting laser), is being developed to allow gigabit transmission over multimode fiber.

Although four graded index multimode fibers have been used over the history of fiber optic communications, one fiber now is by far the most widely used, the 62.5/125. Virtually all multimode datacom networks use this fiber. The first multimode fiber widely used was the 50/125, first by the telephone companies that needed its greater bandwidth for long-distance phone lines. The 50/125 fiber had the highest bandwidth of all multimode fibers, so it is now being reconsidered for specialized high-speed applications. But manufacturers are now introducing higher-bandwidth 62.5/125 fiber that offers the same performance as the 50/125 with the added benefit of compatibility with current installations.

Because the small core and low NA of 50/125 fiber made it difficult to couple to LED sources, many data links switched to 100/140 fiber. The 100/140 fiber worked well with these data links, but its large core made it costly to manufacture, and its unique cladding diameter required connector manufacturers to make connectors specifically for it. These factors led to its declining use. The final multimode fiber, 85/125, was designed by Corning to provide efficient coupling to LED sources but use the same connectors as other fibers. However, once IBM standardized the 62.5/125 fiber for its fiber optic products, the usage of all other multimode fibers declined sharply.

All singlemode links are very high bit rate and long-distance applications, like telephone networks or CATV systems. They use laser sources at either 1,300 nm or 1,500 nm if longer distances are expected. Multimode fiber is used for short links, like LANs or security systems, and at lower bit rates.

The telcos switched to singlemode fiber for its better performance at higher bit rates and its lower loss, allowing faster and longer unrepeatered links for long-distance telecommunications. Virtually all telecom applications use singlemode fiber. It is also used in CATV, since analog CATV networks use laser sources designed for singlemode fiber. Other high-speed networks are using singlemode fiber, either to support gigabit data rates or long-distance links.

## FIBER ATTENUATION

The most important characteristic of the fiber is the attenuation or loss of light as it travels down the fiber. The attenuation of the optical fiber is a result of two factors, absorption and scattering (Figure 11–3). The absorption is caused by the absorption of the light and conversion to heat by molecules in the glass. Primary absorbers are residual OH+ and dopants used to modify the refractive index of the glass. This absorption occurs at discrete wavelengths, determined by the elements absorbing the light. The OH+ absorption is predominant, and occurs most strongly around 1,000 nm, 1,400 nm, and above 1,600 nm.

The largest cause of attenuation is scattering. Scattering occurs when light collides with individual atoms in the glass and is anisotrophic. Light that is scattered at angles outside the numerical aperture of the fiber will be absorbed into the cladding or transmitted back toward the source. Scattering is also a function of wavelength, proportional to the inverse fourth power of the wavelength of the light. Thus if you double the wavelength of the light, you reduce the scattering losses by twenty-four or sixteen times. Therefore, for long-distance transmission, it is advantageous to use the longest practical wavelength for minimal attenuation and maximum distance between repeaters. Together, absorption and scattering produce the attenuation curve for a typical glass optical fiber shown in Figure 11–3.

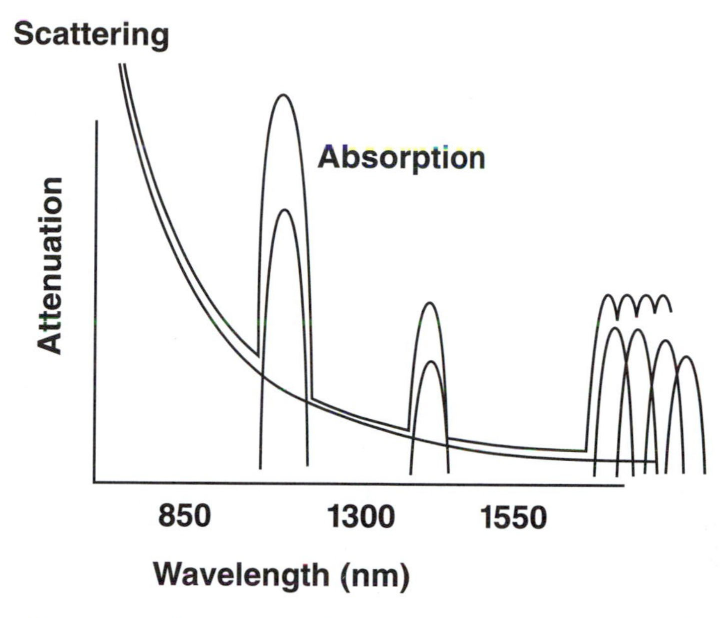


**Figure 11–3.** Fiber attenuation is caused by a combination of scattering and absorption.

Fiber optic systems transmit in the "windows" created between the absorption bands at 850 nm, 1,300 nm, and 1,550 nm, where physics also allows one to fabricate lasers and detectors easily. Plastic fiber has a more limited wavelength band that limits practical use to 660-nm LED sources.

## FIBER BANDWIDTH

The information transmission capacity of fiber is limited by two separate components of dispersion: modal and chromatic. Modal dispersion (Figure 11–4) occurs in step index multimode fiber where the paths of different modes are of varying lengths. Modal dispersion also comes from the fact that the index profile of graded index (GI) multimode fiber is not perfect. The graded index profile was chosen to theoretically allow all modes to have the same group velocity or transit times.

By making the outer parts of the core a lower index of refraction than the inner parts of the core, the higher-order modes speed up as they go away from the center of the core, compensating for their longer path lengths. The index of refraction is a measure of the speed of light in the glass, so the light actually travels faster farther from the core of the fiber. This effect causes the higher-order modes to follow a curved path that is longer than the axial ray (the "zero-order mode"), but by virtue of the lower index of refraction away from the axis, light speeds up as it approach-

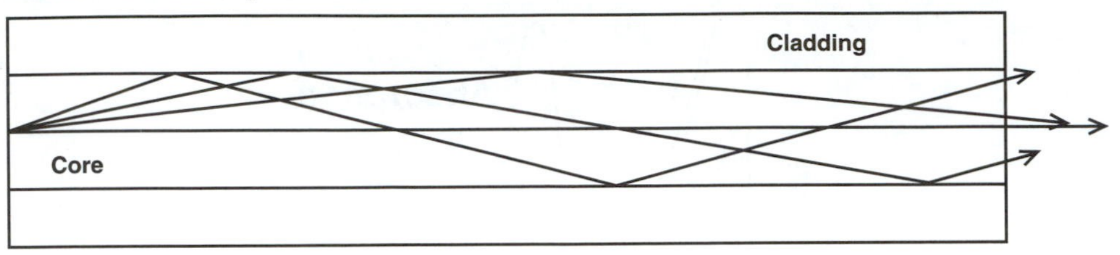


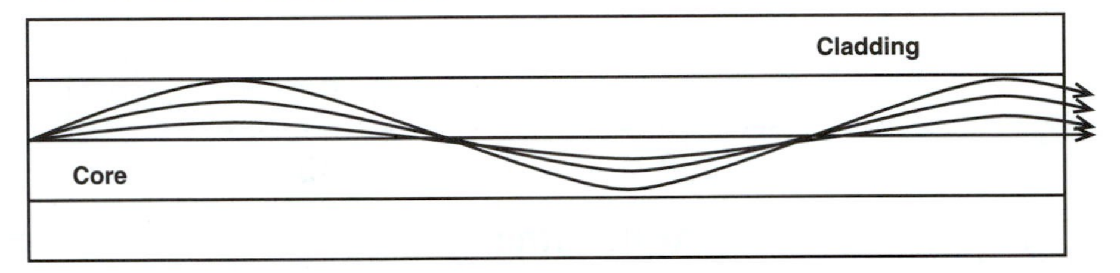


**Figure 11–4.** Modal dispersion caused by different path lengths in the core of the fiber is one aspect of fiber bandwidth.

es the cladding and takes approximately the same time to travel through the fiber. Thus the "dispersion" or variations in transit time for various modes is minimized and the bandwidth of the fiber is maximized.

Chromatic dispersion is caused by the light of different wavelengths traveling at different speeds. The index of refraction of glass is also a function of the wavelength of light, as is shown by the dispersion of sunlight into a spectrum by a prism. The wavelength and spectral width of the LED or laser source can affect the bandwidth, since longer-wavelength light travels faster through glass (Figure 11–5), so the light will be dispersed when traveling through the fiber. The amount of dispersion is determined by the characteristics of the glass in the core of the fiber and the spectral characteristics of the source. LEDs are comprised of more wavelengths of light and therefore are more affected by chromatic dispersion than lasers.

Singlemode fiber does not have modal dispersion, but its ultimate bandwidth is limited by the spectral characteristics of the laser source. Thus the actual bandwidth of a fiber is determined by the characteristics of the fiber itself and the source used in the data link. Link designers must consider all these factors when designing systems that use multimode fiber.

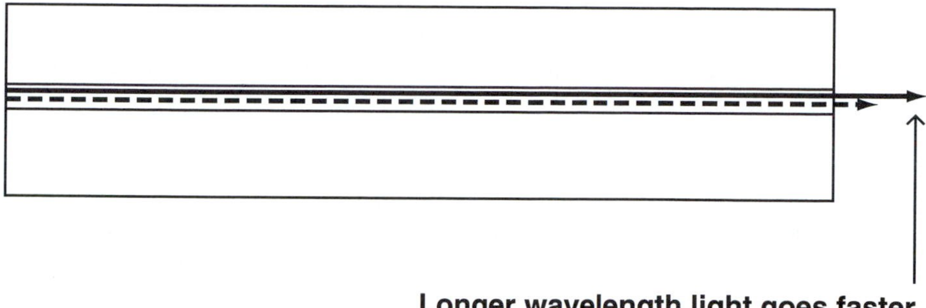


**Figure 11–5.** Chromatic dispersion occurs because light of different colors (wavelengths) travels at different speeds in the core of the fiber.

## BENDING LOSSES

Fiber is subject to additional losses as a result of stress. In fact, fiber makes a very good stress sensor. fiber optic cables are specifically designed to prevent fiber from being stressed or damaged by the environment in which it is installed. It is also mandatory to minimize stress and/or stress changes on the fiber when manufacturing the cable, installing it, and making measurements.

## FIBER OPTIC CABLE

The main role of fiber optic cable is to protect the fiber. Cable comes in many different types, depending on the number and types of fibers and the environment where it will be installed. One must choose fiber optic cable carefully, as the choice will affect how easy it is to install, splice, or terminate, and most important, what it costs.

### Choosing a Cable

Since the job of the cable is to protect the fibers from the hazards encountered in an installation, there are many types from which to choose. Cable choice depends on where the cable will be run. Inside buildings, cables do not have to be as strong to protect the fibers, but they have to meet all fire code provisions. Outside buildings, cable type depends on whether the cable is buried directly, put in conduit, strung aerially, or even placed underwater.

The best source of cable information is cable manufacturers. Contact several of them (two minimum, three preferred) and give them the details of the installation. They will want to know where the cable is going, how many fibers you need, and what kind you need (singlemode or multimode, or both in hybrid cables). Some cables have metal strength members or even metal signal or power cables; they are called composite cables. The cable companies will evaluate your requirements and make suggestions, and then you can get competitive bids.

Since the application will call for a certain number of fibers, consider adding spare fibers to the cable. That way, spares will be available if you break a fiber or two when splicing, breaking out, or terminating fibers. And always consider future expansion. Most users install many more fibers than needed, especially by adding singlemode fiber to multimode fiber cables for campus or backbone applications. It is not uncommon to install more than twice as many fibers as needed to allow for future expansion.

### Cable Types

All cables share some common characteristics. They all include various plastic coatings to protect the fiber, from the buffer coating on the fiber itself to the outside jacket. All include some strength members, usually a high-strength "aramid" yarn often called "Kevlar," which is the duPont trade name, to use in pulling the cable without harming the fibers. Larger cables with more fibers usually have a fiberglass rod down the middle for more strength and to limit the bend radius. The following are the standard cable types, although the cable makers sometimes have slightly different names for them.

*Simplex cable and zip cord (Figure 11–6).* A simplex cable consists of one fiber, with a 900 micron buffer coating, Kevlar strength member and PVC jacket. The jacket is usually 3 mm (1/8 inch) diameter. Zip cord is simply two of these cables joined with a thin web. It is used mostly for patchcord and backplane applications, but zip cord can also be used for desktop connections.

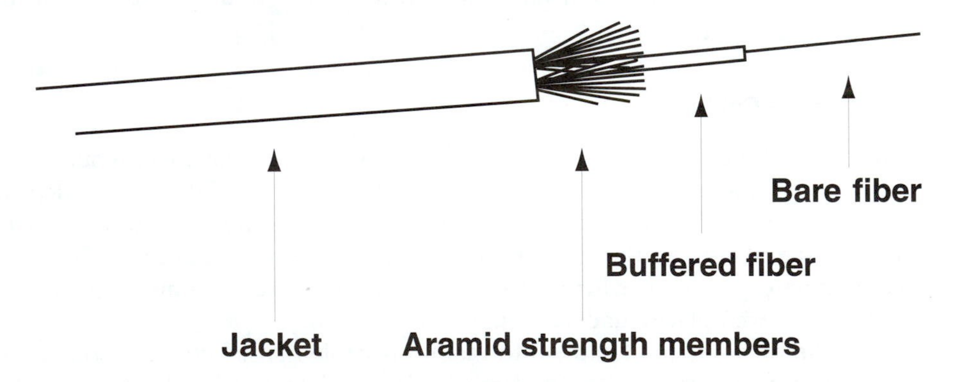


**Figure 11–6.** Simplex cable has only one fiber.

*Distribution cables (Figure 11–7).* They contain several 900 micron-buffered fibers bundled under the same jacket with Kevlar or fiberglass rod reinforcement. These cables are small in size and used for short, dry conduit runs, and for riser and plenum applications. The fibers are double-buffered and can be directly terminated, but because their fibers are not individually reinforced, these cables need to be broken out with a "breakout box" or terminated inside a patch panel or junction box.

*Breakout cables (Figure 11–8).* These cables are made of several simplex cables bundled together. This is a strong, rugged design, but they are larger and more expensive than the tightpack or distribution cables. This cable is suitable for conduit runs and for riser and plenum applications. Because each fiber is individually reinforced, this design allows for quick termination to connectors. Breakout cable can be more economical where fiber count is not too large and distances are not too long, because it requires so much less labor to terminate.

*Loose tube cables (Figure 11–9).* These cables are composed of several fibers together inside a small plastic tube, which are in turn wound around a central strength member and jacketed, providing a small, high fiber count cable. This type of cable is ideal for outside plant trunking applications, as it can be made with the loose tubes filled with gel to prevent harm to the fibers from water. It can be used in conduits, strung overhead, or buried directly into the ground. Since the fibers have only a thin buffer coating, they must be carefully handled and protected to prevent damage.

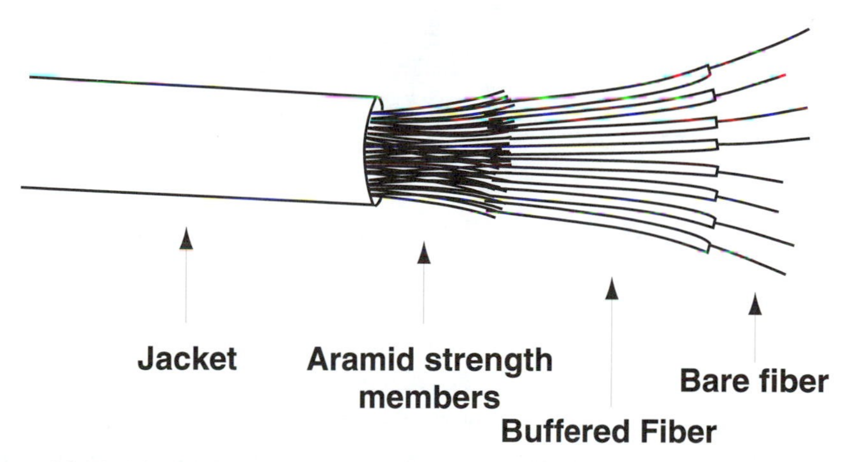


**Figure 11–7.** Distribution cables are similar to simplex cables, except they have several fibers in the center of the cable.

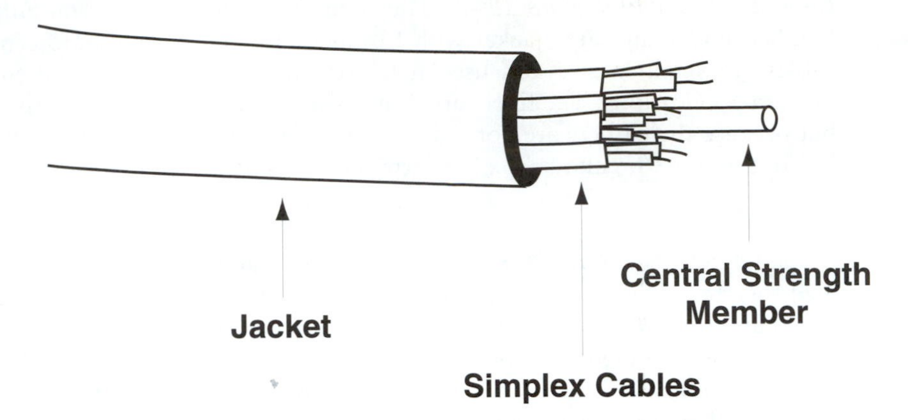


**Figure 11–8.** Breakout cables are a number of simplex cables wound around a central strength member and jacketed.

*Other Types.* There are other cable types like ribbon cable, and there are different names given to the types already discussed. Every manufacturer has its own favorites, so it is a good idea to get literature from as many cable makers as possible. And do not overlook the smaller manufacturers—often they can help you save costs by making special cable for you.

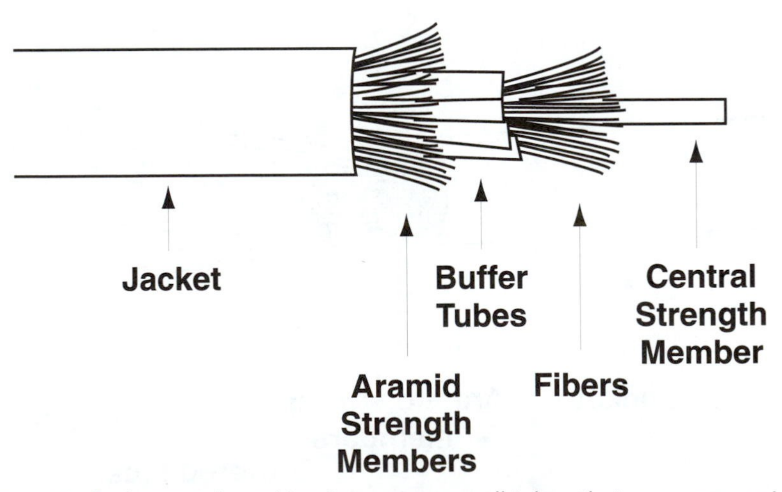


**Figure 11–9.** Loose tube cables have many small tubes that can carry up to 12 fibers each.

### Cable Ratings and Markings

All cables must carry identification and ratings per the *NEC®* (*National Electrical Code®*) Chapter 770. Cables without markings should never be installed, as they will not pass inspections. The ratings are:

| | |
|---|---|
| OFN | optical fiber nonconductive |
| OFC | optical fiber conductive |
| OFNG or OFCG | general purpose |
| OFNR or OFCR | riser rated cable for vertical runs |
| OFNP or OFCP | plenum rated cables for use in air-handling areas |
| OFN-LS | low smoke density |

## FIBER OPTIC CABLE INSTALLATION

Installing fiber optic cable is simplified by the hardy construction of the cable itself. Although the fiber is usually glass, which is perceived as fragile, it is actually stronger than steel. However, if it is bent over too tight a radius, especially under tension, it may fracture. Manufacturers of cable, therefore, design the cable to protect the fiber under stress. While Cat 5 UTP is limited to a 25-pound pulling tension, most indoor fiber optic cables are rated at over 100 pounds, and outside plant cables may be rated to 600 pounds or more.

All fiber optic cables must be pulled by the strength members unless the cable has been specifically designed to be pulled by the jacket. Most cables are designed to be pulled by the strength members included in the cable. Typically, duPont Kevlar® or another aramid fiber will be included as a strength member for pulling. In preparation for pulling, the cable jacket should be stripped, fibers and any internal stiffeners cut off, and the pulling eye attached to the strength members only.

Never pull a cable by the fibers or harm will be done to them. Pulling fibers by the jacket usually results in the cable stretching under tension, then retracting, causing the fibers to be put under great stress. Only specialized cables with double jackets or armoring can normally be safely pulled by the jacket.

Under circumstances where the tension is not too large, small fiber count cables can be pulled by wrapping several turns around a large diameter mandrel to distribute the tension along a length of the cable to reduce the stress on any part of the jacket. The cable must not overlap on the mandrel and must be kept snugly wound on the mandrel. Fiber spools make excellent mandrels for pulling smaller cables.

While fiber optic cable can withstand great tension, it still requires care in installation. Twisting the cable is potentially harmful, so it should be unspooled by rolling directly off the reel, not off the ends. If it is unspooled for pulling, it can be laid on the ground in a "figure 8" pattern, which prevents twisting. Even when pulling, a swivel eye should be used to prevent twisting from the pulling rope or tape.

If the fiber is to be pulled around a corner, care should be taken to minimize both the pulling tension and bending radius. Observe the manufacturers' recommendations, or if they are not known, assume a bending radius under tension of twenty times the cable diameter.

Fiber optic cable should not be left unsupported, nor should it be covered by heavier copper cables. If cable trays are used, fiber should be installed last, on top of the copper cables or lashed to the bottom of the tray. If "J hooks" are used with fiber, use the wide ones specifically designed for Cat 5.

Outside plant cable can be direct buried or pulled in conduit. Long cable runs, up to several kilometers, can be installed with proper cable lubrication and pullers that monitor and limit tension. Lubricants should be chosen for compatibility to the cable jacket to prevent long-term damage to the cable. If a single run is desired but it is too long or has too many bends for a single pull, the cable can be pulled from an intermediate point, despooled into a figure 8, and pulled in the opposite direction.

Aerial installations will require a self-supporting cable or attachment to a messenger. All dielectric aerial cable is available, as well as cable with an attached metal messenger for support, but many outside cables can be lashed for aerial installation.

It is wise to try to pull cable between locations without splicing, to reduce installation complexity and cost. The cost is not only the cost of making a splice but also that of making the enclosures and providing proper mounting for the enclosures. Connectors and patch panels or boxes will prove to be more cost-effective in most installations.

Fiber optic cable is typically installed, then terminated on-site. Cable can sometimes be installed with connectors already installed if proper precautions are taken to protect the connectors. A protective boot must be installed over the connectors and attached to the strength members of the cable before pulling. Since this makes the cable much more bulky, pulling becomes more difficult.

Two other alternatives to field termination are available. You can terminate one end of the cable then pull the unterminated end, reducing the number of field terminations by one-half. Or you might consider some of the new multifiber connectors that have up to twelve fibers in a connector that is smaller than some single fiber connectors.

Whenever pulling preterminated cable, remember that accurate length calculations are mandatory to prevent wasting cable or being too short. Consider vertical runs (such as from a floor outlet to above a ceiling) and service loops as part of the length.

In fact, service loops should be included on all cable installations. This extra length could be critical if splicing for restoration or retermination ever becomes necessary. Coil up the excess fiber where it will not be harmed but where its location will be obvious when it is needed.

## CHAPTER REVIEW

1. Label the following on the diagram: core, cladding, and buffer.

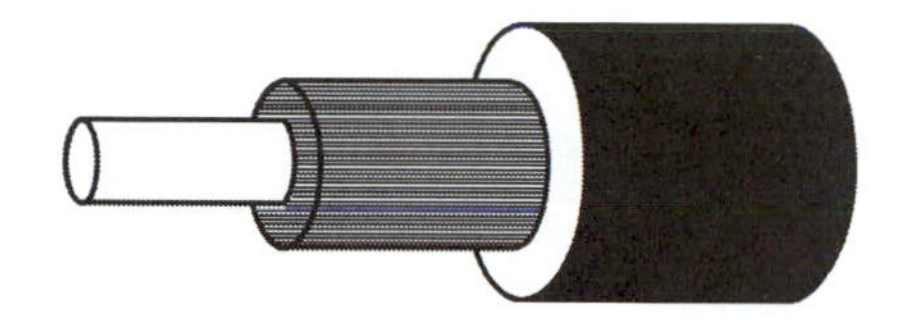

2. What part of the fiber carries the light? What part traps light in the core?
3. What is defined by the highest angle at which the light is accepted into the core of the fiber?
4. In multimode fiber, does the light travel faster near the center or the outside of the core?
5. How many times more is the bandwidth capacity of a multimode graded index cable than a step index cable?
6. What fiber can support gigabit data rates over long-distances?
7. What two factors cause attenuation?
8. In what wavelength windows do fiber optic systems transmit?
9. List two types of dispersion.
10. What is the main purpose of the fiber optic cable?
11. What standards must inside cables meet?
12. What does a hybrid cable have?
13. What do you use to pull the fiber cable?
14. What cable is used in an environmental air area?
15. Which fiber cable is easy to terminate?
16. The cable that has a gel to prevent water getting to the fiber is called what?
17. Name the cable rating or marking of the fiber cable used in vertical runs.
18. When you do not know the bend radius, what should you use as a "rule of thumb"?
19. How may you accomplish a long single pull run?
20. How can you reduce the number of field terminations?

CHAPTER

# 12

# FIBER OPTIC CONNECTORS AND SPLICES

Fiber optic connectors are used to couple two fibers together or to connect fibers to transmitters or receivers. Connectors are designed to be demountable. Splices, however, are used to connect two fibers in a permanent joint. While they share some common requirements, like low loss, high optical return loss, and repeatability, connectors have the additional requirements of durability under repeated matings. Splices, meanwhile, are expected to last for many years through sometimes difficult environmental conditions.

## CONNECTORS

Since the beginnings of fiber optics, there have been approximately seventy different connectors in use. Figure 12–1 shows the connector types that are most widely used today, although the SMA and Biconic are becoming obsolete. Most connectors work by simply aligning the two fiber ends as accurately as possible and securing them in a fashion that is least affected by environmental factors. The most common method is to have a cylindrical ferrule with a fiber-sized hole in the center, in which the fiber is secured with an adhesive. Note that fiber optic connectors are mainly "male" style with a protruding ferrule, since the end of the ferrule must be polished after the fiber is glued into it.

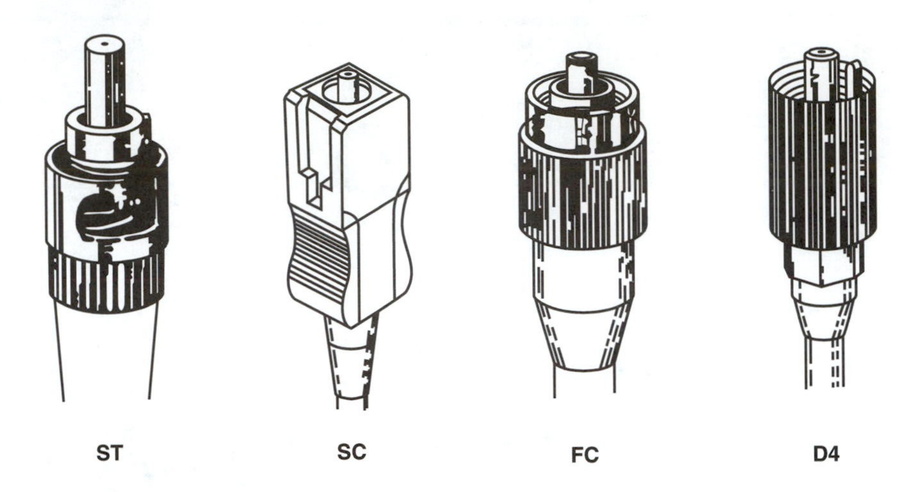


**Figure 12–1.** The most common fiber optic connectors.

Many other connector techniques like expanded beam, use of lenses, and alignment of bare fibers like in a splice have been tried, but most have been abandoned for all but some very specialized applications.

Connectors have used metal, glass, plastic, and ceramic ferrules to align the fibers accurately, but ceramic seems to be the best choice. It is the most environmentally stable material, closely matching the expansion coefficient of glass fibers. It is easy to bond to glass fiber with adhesives, and its hardness is perfect for a quick polish of the fiber. As volume has increased, ceramic costs have been reduced to be competitive with the cost of metal connectors. A new type of plastic, liquid crystal polymers (LCP), offers promise for molded ferrules at lower costs, once performance and durability are proven.

Splice bushings used to align two connectors have been made from metal, plastic, and ceramic also. The plastic types work well over environmental conditions but suffer degradation with repeated matings, especially under conditions encountered in testing. The plastic bushings "shave" small amounts of plastic at each insertion. Some of this material may accumulate on the end of the connector and cause loss. Some may also build up and form a ridge in the bushing to cause an end gap in the mating of two connectors. Check splice bushings for these problems by viewing the end of the connector in a microscope and looking for dust.

### Most Popular Connector Types

The connectors shown in Figure 12–1 are the most common fiber optic connectors. The ST is by far the most popular multimode connector because it is cheap and easy to install. The SC connector was specified as a standard by the EIA/TIA 568 specification, but its higher cost and difficulty of installation limited its popularity until recently. However, newer SCs are much better in both cost and installation ease, so its popularity has been growing rapidly; it now represents over 15 percent of the market. The duplex FDDI, ESCON, and SC connectors are used for patchcords to equipment and can be mated to ST or SC connectors at wall outlets.

Singlemode networks use FC or SC connectors in about the same proportion as ST and SC are used in multimode installations. STs are used in smaller numbers. There are also some D4s and Biconic connectors still in use.

But the connector scene is changing. There are several new small duplex connectors being introduced (Figure 12–2), including the tiny Lucent LC, the Panduit "Opti-Jack" that is an optical RJ-45, the 3M "Volition," the MT-RJ supported by numerous manufacturers, and the SC-DC and SC-QC from Siemens and IBM. These connectors offer much higher density of fiber optic interfaces in patch panels and networking equipment, so they are likely to become very popular. The standards committees have declined to choose a new standard, so the marketplace must decide which of these connectors to use.

### Termination Procedures

Multimode connectors are usually installed in the field on the cables after pulling, while singlemode connectors are usually installed by splicing a factory-made "pigtail" onto the fiber. That is because the tolerances on singlemode terminations are much tighter and the polishing processes are more critical, making consistent field termination difficult. In the factory, special polishing techniques are used to minimize reflections and singlemode terminations also. One can install singlemode connectors in the field for data networks, but losses may be 1 dB.

Cables can be pulled with connectors already on them if you can deal with the following two problems. First, the length must be precise. Too short, and you will have to repull another longer one (it is not cost-effective to splice); too long, and you

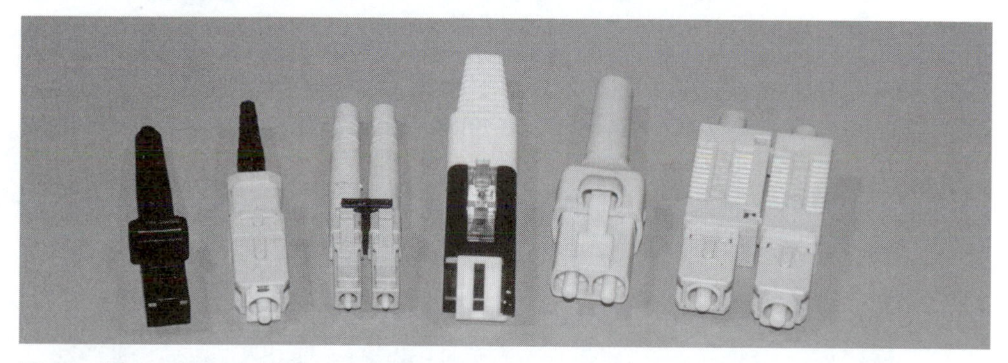

**Figure 12–2.** New small form factor connectors, from left, MT-RJ, SC/DC, LC, Volition™, Opti-Jack, and the current Duplex SC.

waste money. Second, the connectors must be protected. Some cable and connector manufacturers offer protective sleeves to cover the connectors, but you must still be much more careful in pulling cables. You might consider terminating one end and pulling the unterminated end in order to avoid putting the connectors at risk.

### Termination Methods

*Multimode termination.* Several different types of terminations are available for multimode fibers. Each version has its advantages and disadvantages, so learning more about how each works will help you to decide which one to use. Most connectors use epoxies or other adhesives to hold the fiber in the connector. It is critical to use only the specified epoxy, as the fiber-to-ferrule bond is critical for low loss and long term reliability.

All termination methods have a similar process and use similar tools. Figure 12–3 shows a termination setup for epoxy polish connectors. The typical termination process includes these steps:

1. Place a strain relief and crimp bushing on the fiber.
2. Strip the jacket off the cable to expose about 1 1/2 inches of fiber.
3. Cut the strength members to approximately 3/8 inches in length.

**Figure 12–3.** Terminating fiber optic cable with epoxy polish connectors.

4. Strip the buffer to expose 3/4 inches of bare fiber.
5. Clean the fiber with an alcohol-saturated lint-free wipe.
6. Inject adhesive into the connector until a bead forms on the end of the ferrule.
7. Insert the fiber fully.
8. Capture the strength members under the crimp bushing and crimp.
9. Slide the strain relief over the crimp bushing.
10. Let the adhesive set, or cure it in an oven.
11. After curing, cleave the fiber at the end of the ferrule.
12. Polish off the exposed fiber and epoxy by "air polishing" with a relatively coarse 15-mm film.
13. Polish with a medium 3-µm film until it feels "slick."
14. Final polish with 0.3-µm film.
15. Inspect your work with a microscope.

This entire process can be done in five minutes with experience, and the yield is over 95 pecent with epoxy/polish connectors.

*Epoxy/Polish.* Most connectors are the simple "epoxy/polish" type where the fiber is glued into the connector with epoxy and the ferrule end is polished with special polishing film. These connectors provide the most reliable connection, the lowest losses (less than 0.5 dB), and the lowest costs, especially if you are doing a large number of connectors. The epoxy can be allowed to set overnight or cured in an inexpensive oven in only a few minutes (Figure 12–4). A "heat gun" should never be used to try to cure the epoxy faster, as the uneven heat may not cure all the epoxy or may overheat some of it and prevent it from ever curing!

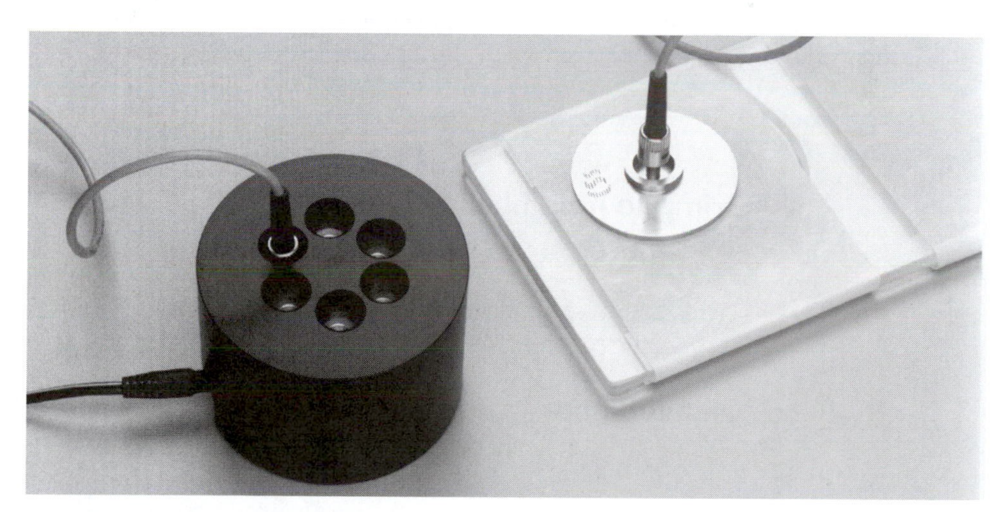

**Figure 12–4.** Portable connector curing oven.

*"Hot Melt."* This is a 3M trademark for a connector that has the adhesive inside the connector. You heat the connector to a high temperature and insert the fiber while the adhesive is still hot and liquid. The adhesive sets as the connector cools, and it is then polished like an epoxy/polish connector. Since the hot melt is fast to teminate but has a high cool, it is used by installers who only have small quantities of connectors to terminate.

*Crimp/polish.* Rather than glue the fiber in the connector, these connectors use a crimp on the fiber to hold it in place. Early types offered "iffy" performance, but today these connectors work quite well, if you practice a lot with them. Expect to trade higher losses for the faster termination speed. And they are more costly than epoxy/polish types. These connectors are a good choice if you only install small quantities and your customer will accept the higher losses expected of them.

*Quick-setting adhesives.* These connectors use a quick-setting adhesive with a curing agent to replace the epoxy. They work well if your technique is good and you work fast. If you use improper techniques, you will have fibers halfway into the connector and solidly glued in place, and the connector will have to be discarded. Practice is important if these connectors are to be used. Also, some adhesives may not be qualified for broad temperature ranges or long-term reliability, unless specifically stated by the manufacturer.

*Prepolished/splice (Figure 12–5).* Some manufacturers offer connectors that have a short stub fiber already epoxied into the ferrule and polished perfectly, so you just cleave a fiber and insert it like a splice. (See the next section on splicing.) Although this technique seems like a great idea, it has several disadvantages. First, it is very

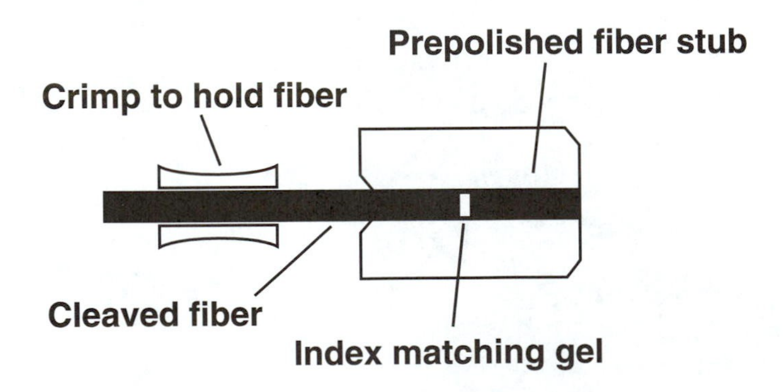


**Figure 12–5.** Prepolished connectors have a short fiber piece already glued in the ferrule.

costly, five to ten times as much as an epoxy/polish type. Second, you have to make a good cleave to make the fibers low loss, and that is not easy. Third, even if you do everything correctly, your loss will still be higher, because you have a connector loss plus two splice losses at every connection. Finally, these connectors have considerably lower yield in termination. These connectors may be good for quick restoration, but look at the cost carefully before you commit to a job with them, and practice with them before using them in the field.

### Hints for Doing Field Terminations

Here are a few things to remember when you are terminating connectors in the field. Following these guidelines will save you time, money, and frustration:

1. Choose the connector carefully, and clear it with the customer if it is anything other than an epoxy/polish type. Some customers have strong opinions on the types or brands of connectors to be used in the job. Find out first, not later!
2. Never, never, NEVER take a new connector in the field until you have installed enough of them in the office that you can put them on in your sleep. The field is no place to experiment or learn. Doing so can cost you big time!
3. Have the right tools for the job. Make sure that you have the proper tools and that they are in good shape before you head out for the job. This includes all the termination tools, cable tools, and test equipment. More and more installers are owning their own tools, like auto mechanics—they say that is the only way to make sure the tools are properly cared for.
4. Dust and dirt are your enemies. It is very hard to terminate or splice in a dusty place. Try to work in the cleanest possible location. Use lint-free wipes (not cotton swabs or rags made from old T-shirts) to clean every connector before connecting or testing it. Do not work under heating vents, as they blow dirt down on you continuously.
5. Do not overpolish. Contrary to common belief, too much polishing is just as bad as too little. The ceramic ferrule in most of today's connectors is much harder than the glass fiber. Polish too much, and you create a concave fiber surface, increasing the loss. A few "figure 8s" will be sufficient.
6. Remember, singlemode fiber requires different connectors and polishing techniques. Most SM fiber is terminated by splicing on a preterminated pigtail, but you can put SM connectors on in the field if you know what you are doing. Since you will have much higher loss, approaching 1 dB, and high back reflections, do not try it for anything but data networks, not telco or CATV.

7. Change polishing film regularly. Polishing builds up residue and dirt on the film that can cause problems after too many connectors and cause poor end finish. Check the manufacturers' specs.
8. Put covers on connectors and patch panels when they are not in use. Keep them covered to keep them clean.
9. Inspect and test, then document. It is very hard to troubleshoot cables when you do not know how long they are, where they go, or how they tested originally! So keep good records. Smart users will require it and will expect to pay extra for good records.

## SPLICES

Splicing is needed only if the cable runs are too long for one straight pull or you need to mix a number of different types of cables (for example, you need to splice a 48-fiber cable to six 8-fiber cables, although in such a case a breakout cable might have worked instead). And, of course, we use splices for restoration after the number one problem of outside plant cables—a dig-up and cut of a buried cable (usually referred to as "backhoe fade" for obvious reasons)—occurs!

There are two types of splices, fusion and mechanical (Figure 12–6). Fusion splicing is done by welding the two fibers together, usually with an electrical arc with an automated splicer. It has the advantages of low loss, high strength, low back reflection (optical return loss), and long-term reliability. Mechanical splices use an alignment fixture to mate the fibers and either a matching gel or epoxy to minimize back reflection. Some mechanical splices use bare fibers in an alignment bushing,

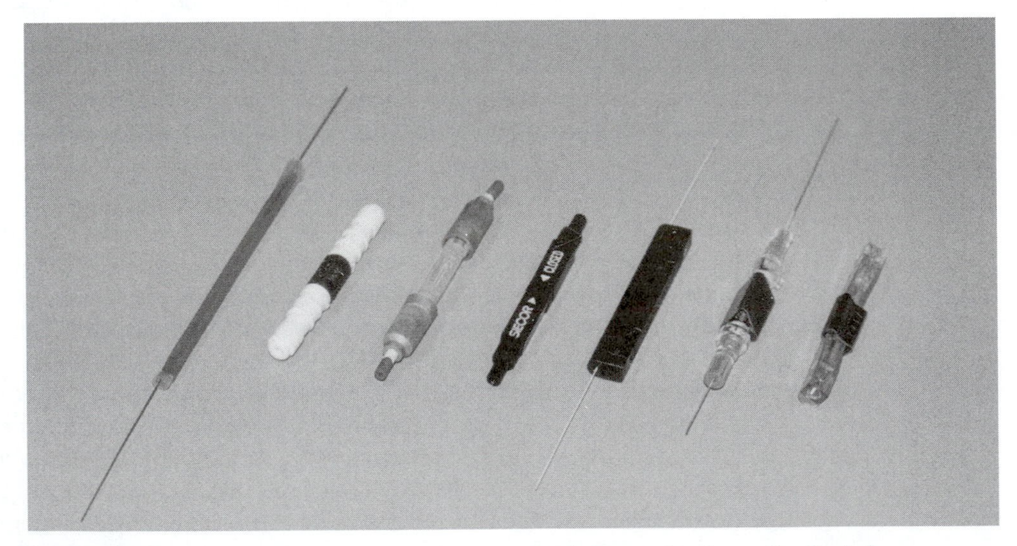


**Figure 12–6.** Fusion and mechanical splices; the fusion splice is on the far left.

while others closely resemble connector ferrules without all the mounting hardware. While fusion splicing normally uses active alignment by the fusion splicing equipment to minimize splice loss, mechanical splicing relies on tight dimensional tolerances in the fibers to minimize loss.

Low splice loss and high return loss is highly dependent on the quality of the cleave on both fibers being spliced. Cleaving is done by using a sharp blade to put a surface defect on the fiber, then pulling carefully to allow a crack to propagate across the fiber. In order to get good fusion splices, both fiber ends need to be close to perpendicular to the fiber axis. Then, when the fibers are fused, they will weld together properly.

With a mechanical splice, the fibers are pushed together with an index-matching gel or epoxy between them. Since the index matching is not perfect, some reflection may occur. If the fibers are cleaved at an angle, about 8 degrees being best, the reflected light will be absorbed in the cladding, reducing the back reflections. Special cleavers have been designed to provide angle cleaves.

### Which Splice Is Best?

The best splice for an application will be determined by three factors: performance, reliability, and cost.

Fusion splices give lower loss and very low back reflections, so they are preferred for singlemode high-speed digital or CATV networks. However, they do not work as well on multimode fiber, because of the multiple layers in the core and the large size of the multimode fiber core may fuse unevenly, so mechanical splices are often preferred for multimode applications.

If the splice is in a location where failure would be difficult to fix, such as buried underground or underwater, or in a difficult environment, fusion splicing will probably be more reliable, since it simply welds the fibers together and has less likelihood of failure.

If cost is an issue, the answer will depend on the number of splices to be made. Fusion splicing requires expensive equipment, but each splice is inexpensive. Mechanical splicing uses inexpensive tools, but each splice is expensive. If you make a lot of splices (like thousands in a large big telco or CATV network), use fusion splices. If you need just a few, use mechanical splices.

## CONNECTOR AND SPLICE PERFORMANCE

Connector and splice loss is caused by a number of factors, shown in Figure 12–7. Loss is minimized when the two fiber cores are perfectly aligned. Only the light that is coupled into the core of the receiving fiber will propagate, so all the rest of the light becomes the connector or splice loss.

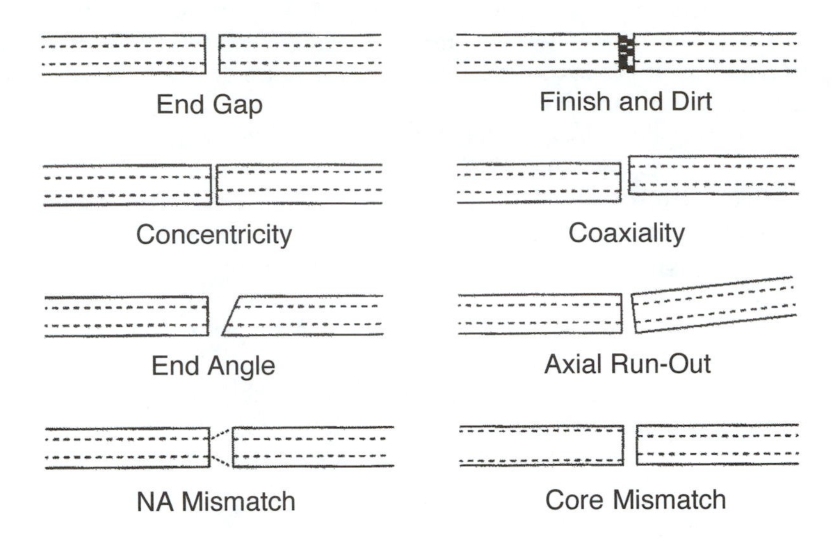


**Figure 12–7.** Causes of loss in fiber optic connectors and splices.

Although it is common to compare the typical connector or splice specifications quoted by manufacturers, such a comparison may not be fair. Each manufacturer has a design that it has qualified by expertly assembling and testing many samples of its connectors. But the actual loss obtained by any end user will be determined primarily by the skill used in the termination process. The manufacturer only has control over the basic design of the connector, the mechanical precision in manufacturing, and the clearness of the termination instructions.

End gaps cause two problems, insertion loss and return loss. The emerging cone of light from the connector will spill over the core of the receiving fiber and be lost. In addition, the air gap between the fibers causes a reflection when the light encounters the change in refractive index from the glass fiber to the air in the gap. This reflection (called fresnel reflection) amounts to about 5 percent in typical flat-polished connectors, and means that no flat-polished connector with an air gap can have less than 0.3 dB loss. This reflection, also referred to as back reflection or optical return loss, can be a problem in laser-based systems. Connectors use a number of polishing techniques to ensure physical contact of the fiber ends to minimize back reflection. On mechanical splices, it is possible to reduce back reflection by using index matching gels and non-perpendicular cleaves, which cause back reflections to be absorbed in the cladding of the fiber.

The end finish of the fiber must be properly polished to minimize loss. A rough surface will scatter light, and dirt can scatter and absorb light. Since the optical fiber is so small, typical airborne dirt can be a major source of loss. Whenever connectors are not terminated, they should be covered to protect the end of the ferrule from dirt. One should never touch the end of the ferrule, since the oils on one's skin causes the fiber to attract dirt. Before connection and testing, it is advisable to clean connectors with lint-free wipes moistened with isopropyl alcohol.

Two sources of loss are directional: numerical aperture (NA) and core diameter. Differences in these two will create connections that have different losses depending on the direction of light propagation. Light from a fiber with a larger NA will be more sensitive to angularity and end gap, so transmission from a fiber of larger NA to one of smaller NA will be higher loss than the reverse. Likewise, light from a larger fiber will have high loss when coupled to a fiber of smaller diameter, while coupling a small diameter fiber to a large diameter fiber produces only minimal loss, since it is much less sensitive to end gap or lateral offset.

These fiber mismatches occur for two reasons: the occasional need to interconnect two dissimilar fibers, and production variances in fibers of the same nominal dimensions. With several multimode fibers in usage today and two others that have been used occasionally in the past, it is possible to sometimes have to connect dissimilar fibers or use systems designed for one fiber on another. Some system manufacturers provide guidelines on using various fibers, but some do not. If you connect a smaller fiber to a larger one, the coupling losses will be minimal, often only the fresnel loss (about 0.3 dB). But connecting larger fibers to smaller ones results in substantial losses, not only due to the smaller core size, but also due to the smaller NA of most small-core fibers.

In Table 12–1, we show the losses incurred in connecting mismatched fibers. The range of values results from the variability of modal conditions. If the transmitting fiber is overfilled or nearer the source, the loss will be higher. If the fiber is near steady state conditions, the loss will be nearer the lower value. If you are connecting fiber directly to a source, the variation in power will be approximately the same as for fiber mismatch, except replacing the smaller fiber with a larger fiber will result in a gain in power roughly equal to the loss in power in coupling from the larger fiber to the smaller one.

**Table 12–1** Mismatched Fiber Connection Losses (excess loss in dB)

| Receiving Fiber | Transmitting Fiber | | |
|---|---|---|---|
| | **62.5/125** | **85/125** | **100/140** |
| 50/125 | 0.9 to 1.6 | 3.0 to 4.6 | 4.7 to 9 |
| 62.5/125 | — | 0.9 | 2.1 to 4.1 |
| 85/125 | — | — | 0.9 to 1.4 |

**Figure 12–8.** Patch panels allow for fiber connections and may have provision for splices also. *Courtesy AMP*

## CABLE PLANT HARDWARE

Fiber optic cable plants require hardware to mount components and protect them. Some fiber optic hardware, such as patch panels (Figure 12–8) and wall outlets, are simple modifications of standard copper wire products used for twisted pair or coax networks. Instead of wire jacks, these products usually have adapters to hold bulkhead splice bushings to mate fiber optic connectors.

When bringing multifiber cables into a building or closet, the cable will likely be terminated in a lockable enclosure to protect the fibers, with access to mating connectors for moves and changes. Similar boxes will be used if splicing cables is necessary.

Outside the building, splicing will be used to join cables. Splice enclosures (Figure 12–9) are available to fit both fusion and mechanical splices and can be direct buried, attached to walls in manholes, or even suspended for aerial installations. Enclosures have trays to organize splices. They may be sealed with gaskets or flooded with compounds to keep water out of the enclosure.

Cables may be installed directly or in conduit, just like copper cables. Many installations use a plastic innerduct even indoors, as it provides protection to the fiber optic cable, even in areas where many other cables are run. Innerduct may also allow faster installation, as it can be run without worry about pulling tension, bending radius, or other obstacles, and then the fiber can be pulled in quickly with little tension (Figure 12–10). Innerduct is available rated for flame retardancy to allow installation in almost any environment.

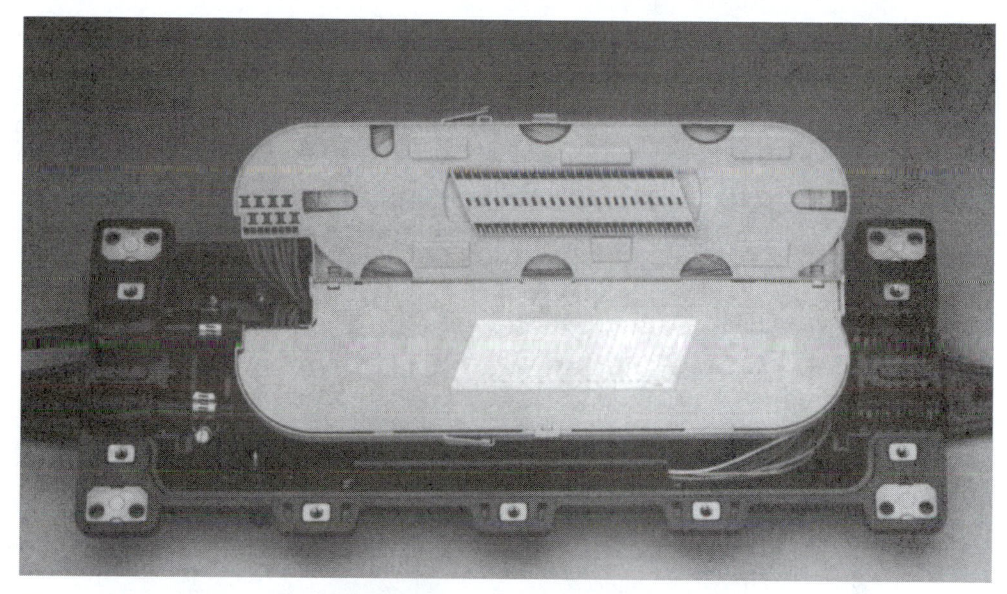

**Figure 12–9.** Outside plant splice closure. *Courtesy 3M*

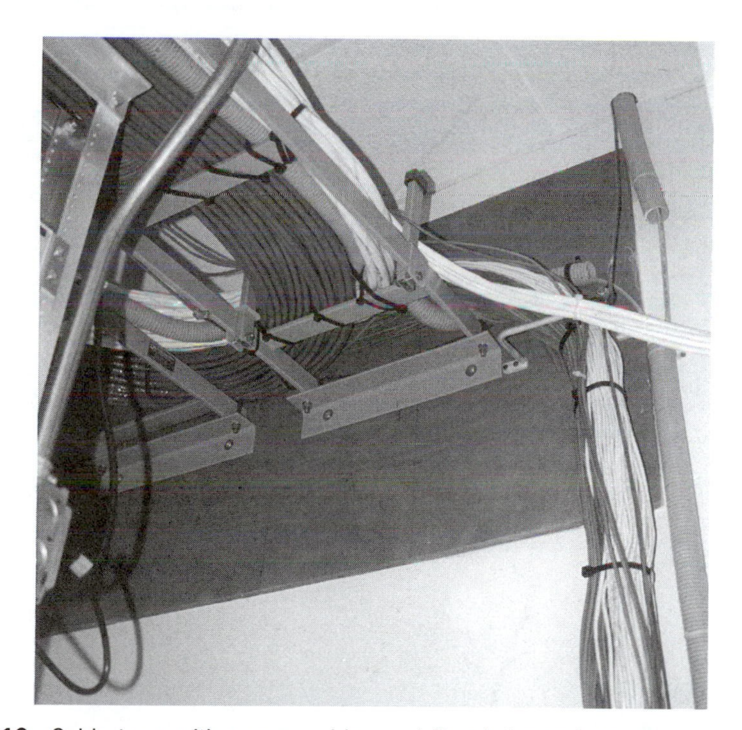

**Figure 12–10.** Cable trays with copper cables and fiber in inner duct.

There are many types of hardware and many manufacturers who make it. Gather catalogs from the hardware makers and/or distributors to see the variety. Look at the places where you will put the hardware, and choose termination boxes or patch panels appropriate for the location. If you can, choose panels that have the connections behind locked doors, since the biggest problem we see is connectors broken at the back end by people working in communications closets. Fiber does not need maintenance or inspection. Lock it up, and unlock it only when you have to move something!

## CHAPTER REVIEW

1. Which is permanent—a splice or a connector?
2. What characteristics are important for both splices and connectors?
3. Why are most fiber optic connectors "male"?
4. What connector ferrule material is used most often?
5. Why are the new small duplex connectors likely to be popular?
6. Why are multimode connectors installed in the field more often than singlemode connectors?
7. Which termination process produces the lowest cost, lowest loss, and most reliable connection?
8. To cure the epoxy, you can leave the connector overnight. Can you use a heat gun for quicker results?
9. Which connection type—Hot Melt, crimp/polish, quick-setting adhesives, prepolished/splice—has two splice losses in addition to a connector loss?
10. When should you try to learn new connectors in the field?
11. What type of work area should one work in with fiber optic connectors?
12. Which splice gives the best long-term reliability?
13. When cleaving, you only put a small defect on the fiber and pull carefully to finish the crack; you never cut through the fiber. T or F
14. How do you minimize the loss of connectors?
15. With mechanical splices, what can you do to reduce back reflection?
16. What do you clean connectors with before connection and testing?
17. What speeds installation and gives better protection to fiber cables?
18. What is the biggest failure problem with connections inside a building?

CHAPTER

# 13

# FIBER OPTIC TESTING

All installed fiber optic networks require testing to ensure the installation was done properly and the cable plant will work with the intended systems. The fiber and cable manufacturer has thoroughly tested the performance of every fiber in every cable. Manufacturers have made thousands of terminations to verify the performance of their splice or connector designs. But the ultimate performance of the cable plant will be determined by the actual installation process, and that can be tested only after the completion of the installation.

Testing fiber optic components and systems requires making several basic measurements. The most common measurement parameters are shown in Table 13–1. Optical power, required for measuring source power, receiver power, and loss or attenuation, is the most important parameter and is required for almost every fiber optic test. Backscatter (OTDR) measurements are the next most important, and bandwidth, wavelength, and dispersion are of lesser importance. Measurement or inspection of geometrical parameters of fiber is essential for fiber manufacturers. And troubleshooting installed cables and networks is required.

## STANDARD TEST PROCEDURES

Most test procedures for fiber optic component specifications have been standardized by national and international standards bodies, including the EIA/TIA in the United States and the IEC internationally. Procedures for measuring absolute opti-

**Table 13–1.** Fiber optic Testing Requirements

| Test Parameter | Instrument |
| --- | --- |
| Optical power (Source output, receiver signal level) | Fiber optic (FO) power meter |
| Attenuation or loss of fibers, cables and connectors | FO power meter and source, test kit, or OLTS (optical loss test set) |
| Source wavelength | FO spectrum analyzer |
| Backscatter (loss, length, fault location) | Optical time domain reflectomete (OTDR) |
| Fault location | OTDR or visual cable fault locator |
| Bandwidth/dispersion (modal and chromatic) | Bandwidth tester or simulation software |

cal power, cable and connector loss, and the effects of many environmental factors (such as temperature, pressure, flexing, and so on) are covered. In performing these tests, the basic fiber optic instruments used are the FO power meter, test source, OTDR, visual fault locator, and connector inspection microscope.

### Measuring Optical Power—Test Instruments

The most basic fiber optic measurement is optical power from the end of a fiber. It is measured with a specialized light meter that has an interface to fiber optic connectors. This measurement is the basis for loss measurements as well as the power from a source or at a receiver. While optical power meters are the primary measurement instrument, optical loss test sets (OLTSs) also measure power in testing loss. EIA standard test FOTP-95 covers the measurement of optical power.

Fiber optic power meters measure optical power with a semiconductor detector that has adapters for different fiber optic connectors. They are calibrated to NIST traceable standards at three primary wavelengths, 850, 1,300 and 1,550 nm. Some meters are also calibrated at 650 nm for plastic optical fiber or 790 nm for the systems that use CD lasers. Meters must also be capable of measuring over a range of optical power appropriate for the systems being tested. Meters appropriate for low-power LED systems, like LANs, may not work properly at the high-power levels of AM CATV links.

### Cable Testing

After the cable has been terminated with connectors or splices, testing will include the loss of the fiber in the cable plus the loss of the connectors. On very short cable assemblies (up to 10 meters long), the loss of the connectors will be the only relevant loss, while fiber will contribute to the overall losses in longer cable assemblies. In an installed cable plant, one must test the entire cable from end to end, including every component in it, such as splices, couplers, connectors, and intermediate patch panels.

The standard cable loss test, FOTP-171, was developed by EIA committees in the mid-1980s and has been used since that time. This test involves launching light from a test source of known power output into the cable and measuring how much power is lost with a fiber optic power meter.

The test equipment includes a test source and fiber optic power meter (Figure 13–1) and reference test cables. The source should be selected to match the type of source (LED or laser) and wavelength used by the system that will operate over the fiber being tested. Test sources have LED or laser diode emitters similar to the ones used in the actual data links. The closer the test source matches the actual system sources, the more accurate loss tests will be. The sources are provided with cou-

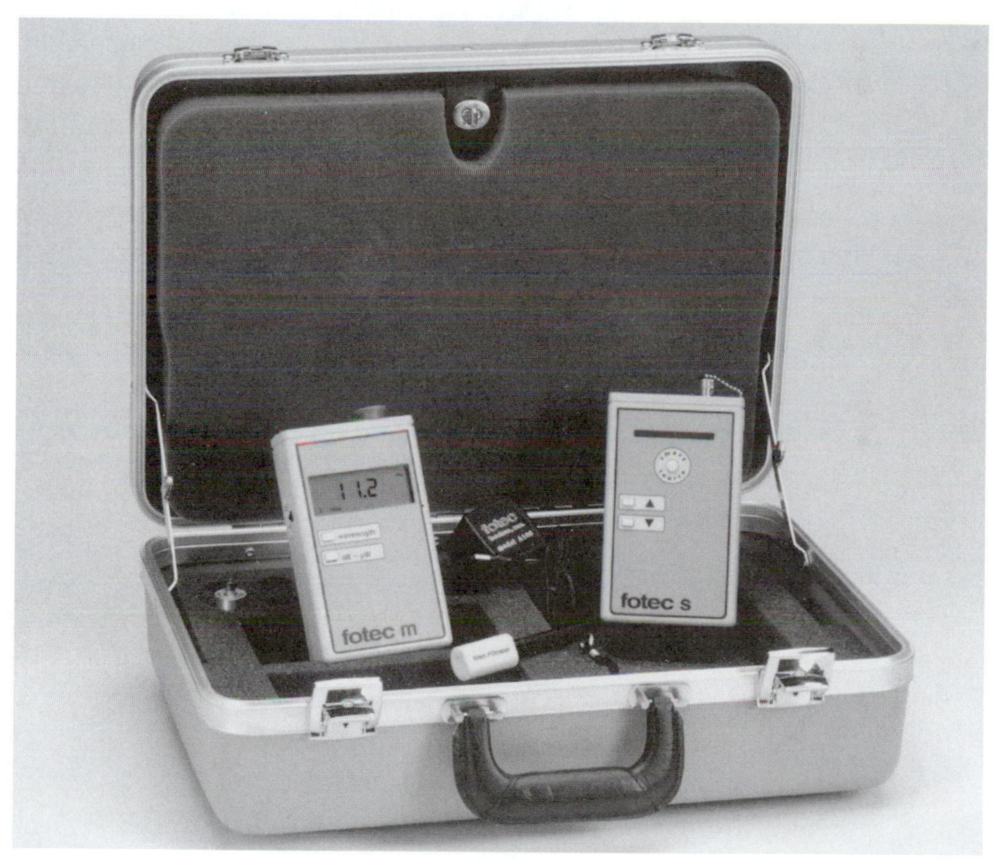


**Figure 13–1.** Fiber optic test kit with source and meter.

plings to standard fiber optic connectors to enable attaching reference cables for loss measurements. The power meter must be chosen to be appropriate to the wavelengths being tested. For field measurements, a resolution of 0.1 dB is appropriate, while lab tests of low loss patchcords may require 0.01 dB resolution.

For multimode fiber systems, testing is usually done at both 850 and 1,300 nm, while singlemode fiber may be tested only at 1,300 nm unless very long cable runs are involved. Testing singlemode fiber at 1,550 nm may also give information on the stress on the cable, as it is much more sensitive to stress at longer wavelengths.

Reference test cables must match the fiber size being tested and have connectors that can be mated to the installed cable. Since many of today's connectors use a common 2.5-mm ferrule, one can use ST connectors to test almost every cable plant by using hybrid mating adapters (ST to SC, ST to FC, ST to FDDI, and so on). The launch cables must be of high quality (that is, low loss connectors) and be kept perfectly clean. Bad or dirty reference cables will result in measurements of high loss, no matter how good the cables being tested may actually be!

The test setup is shown in Figure 13–2. One begins by measuring and/or setting the output of the source "launch cable" to a calibrated value. A launch cable made from the same size fiber and connector type as the cables to be tested is attached to the source. The power from the end of this launch cable is measured by a power meter to calibrate the launch power for the test. If the source is adjustable, it can be set to a convenient reference value, or if the meter has a "zero loss reference" capability, it can be set to read "0 dB" at the output power of the launch cable.

After measuring the reference power, the cable to test is attached to the launch cable and power is measured at the far end. One can calculate the loss incurred in the connector mating to the launch cable and in the fiber in the cable itself.

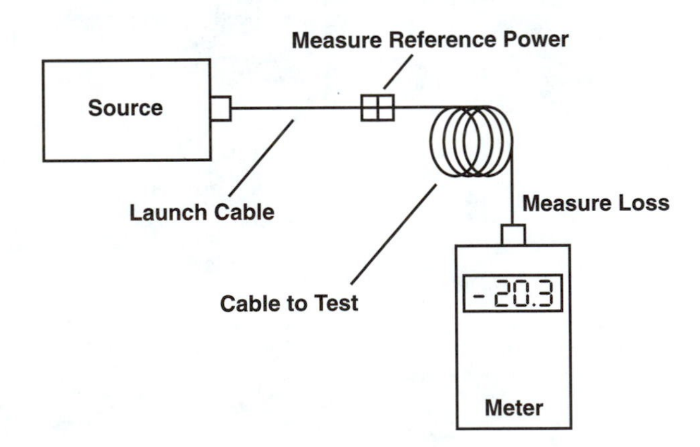


**Figure 13–2.** Single-ended loss testing of a fiber optic cable.

This test only measures the loss in the connector mated to the launch cable, since connecting to the large detector in the power meter effectively has no loss. One can add a second cable at the power meter end, called a receive cable, so the cable to test is between the launch and receive cables (Figure 13–3). Then one measures the loss of both connectors and everything in between. This is commonly called a "double-ended" loss test. The EIA/TIA standard test for testing installed cables, OFSTP-14 (also referred to in EIA/TIA 568 Appendix H), is a double-ended test.

There have been two interpretations of the calibration of the output of the source in this test. In the preferred method, the launch reference cable output power is measured directly by the meter and used as the zero loss reference. This allows one to measure both connectors on the cable being tested, since power is referenced to the output power of the launch connector and there are losses at the connections to both the launch and receive cables.

Another interpretation is that one attaches the launch cable to the source and the receive cable to the meter. The two are then mated and this becomes the "0 dB" reference. With this method, one has two new measurement uncertainties. First, this method underestimates the loss of the cable plant by the loss of one connection, since that is zeroed out in the calibration process. Second, if one has a bad connector on one or both of the test cables, it becomes masked by the calibration. Even if the two connectors have a very high loss, it is zeroed out by the calibration method used.

Obviously, the launch cable reference method is preferred. Both methods are allowed in OFSTP-14, but the launch cable reference will provide lower measurement uncertainty. In addition, one should test the quality of the reference cable connectors by doing a single-ended test with the receive jumper. If this loss is high, one

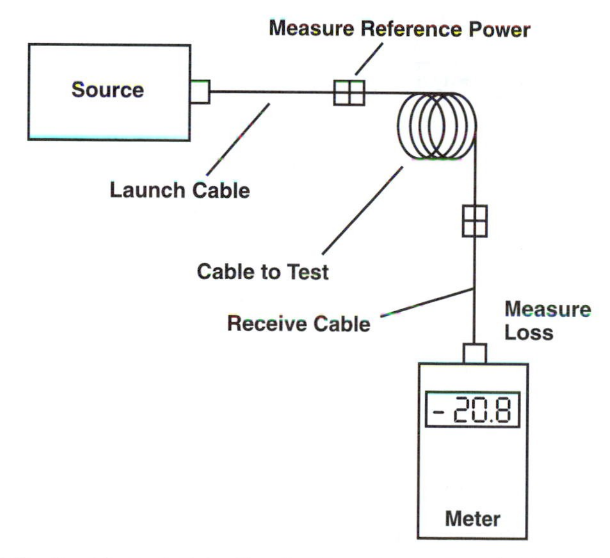


**Figure 13–3.** Double-ended testing tests connectors on both ends of the cables.

knows there is a problem with the test connectors that must be fixed before actual cable loss measurements should be made.

## MEASURING UNITS OF LOSS IN DECIBELS

Loss, the amount of light that is lost in a mated pair of connectors, a splice, or a long length of installed and terminated fiber optic cable, is expressed in "decibels," or "dB." "dB" is a logarithmic scale, dB = 10 log $(P_1/P_2)$, where $P$ is the power in watts. Negative dB (−dB) means a loss in power, such as cable or connector attenuation, while positive dB (+dB) means a gain. Table 13–2 shows the relationship of dB to power ratios.

**Table 13–2**

| dB | Ratio |
|---|---|
| 10 | 10 |
| 20 | 100 |
| 30 | 1000 |
| 3 | 2 |
| 6 | 4 |
| 23=20 + 3 | 100 × 2=200 |

Fortunately, most fiber optic power meters measure directly in dB, so measuring loss is done by simply subtracting power levels. Absolute power, such as the output of a transmitter or the input of a receiver, is expressed in dB relative to 1 milliwatt, or dBm. "0 dBm" equals 1 mW, -10 dBm is 1/10 mW or 100 µW, -20 dBm is 10 µW, and so on.

## WHAT LOSS SHOULD YOU GET WHEN TESTING CABLES?

Although it is difficult to generalize, here are some guidelines:

- For each connector, figure 0.5 dB loss (0.75 maximum).
- For each splice, figure a loss of 0.2 dB.
- For multimode fiber, the loss is about 3 dB/km for 850-nm sources and 1 dB/km for 1,300 nm. This roughly translates into a loss of 0.1 dB per 100 feet for 850 nm, and 0.1 dB per 300 feet for 1,300 nm.
- For singlemode fiber, the loss is about 0.4 dB/km for 1,300-nm sources and 0.3 dB/km for 1,550 nm. This roughly translates into a loss of 0.1 dB per 600 feet for 1,300 nm, and 0.1 dB per 750 feet for 1,300 nm.

For the loss of a cable plant, calculate the approximate loss as:

(0.5 dB × Number of connectors) + (0.2 dB × Number of splices) + Fiber loss on the total length of cable at the specified wavelength

## FINDING BAD CONNECTORS

If a test shows a jumper cable to have high loss, there are several ways to find the problem. If you have a microscope, commonly included in termination kits, inspect the connectors for obvious defects like scratches, cracks, or surface contamination. If they look okay, clean them before retesting. Retest the launch cables to make certain they are good.

Retest the jumper cable with the single-ended method, using only a launch cable. Test the cable in both directions. The cable should have higher loss when tested with the bad connector attached to the launch cable, since the large area detector of the power meter will not be affected as much by the typical loss factors of connectors.

### Inspecting Connectors with a Microscope

Visual inspection of the end surface of a connector is one of the best ways to determine the quality of the termination procedure and to diagnose problems. A well-made connector will have a smooth, polished, scratch-free finish, and the fiber will not show any signs of cracks or pistoning (where the fiber is either protruding from the end of the ferrule or pulling back into it).

The proper magnification for viewing connectors is generally accepted to be 30 to 400 power. Lower magnification, typical with a jeweler's loupe or pocket magnifier, will not provide adequate resolution for judging the finish on the connector. Magnification that is too high tends to make small, ignorable faults look worse than they really are. A better solution is to use medium magnification, about 100x, but inspect the connector three ways at two different angles (Figure 13–4a)—viewing directly at the end of the polished surface with side lighting, viewing directly with side lighting and light transmitted through the core, and viewing at an angle with lighting from the opposite angle (Figure 13-4b).

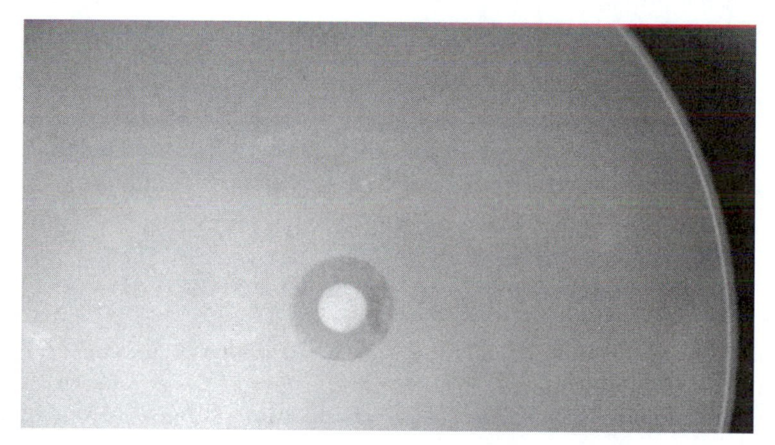

**Figure 13–4a.** Direct view of the polished end of a connector ferrule at 100X.

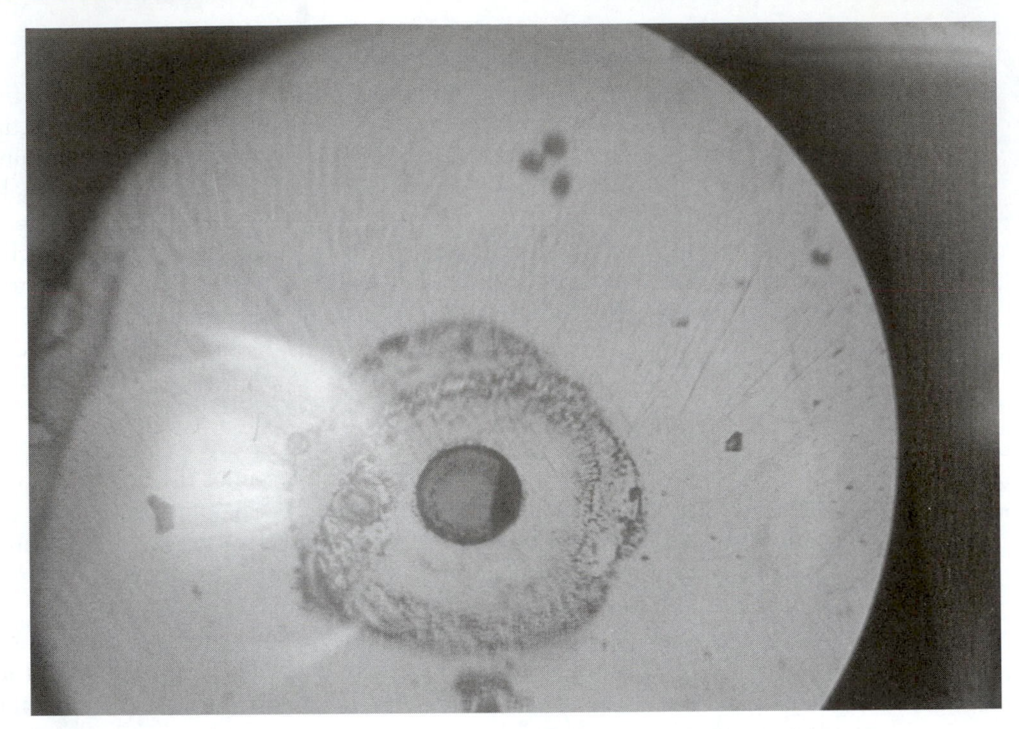

**Figure 13–4b.** Angle view

Viewing directly with side lighting allows you to determine if the ferrule hole is of the proper size, the fiber is centered in the hole, and a proper amount of adhesive has been applied. Only the largest scratches will be visible this way, however. Adding light transmitted through the core will make cracks in the end of the fiber, caused by pressure or heat during the polish process, visible.

Viewing the end of the connector at an angle, while lighting it from the opposite side at approximately the same angle, will allow the best inspection for the quality of polish and possible scratches. The shadowing effect of angular viewing enhances the contrast of scratches against the mirror-smooth polished surface of the glass.

One needs to be careful in inspecting connectors, however. The tendency is to be overly critical, especially at high magnification. Only defects over the fiber core are a problem. Chipping of the glass around the outside of the cladding is not unusual and will have no effect on the ability of the connector to couple light in the core. Likewise, scratches only on the cladding will not cause any loss problems.

### Transmission versus OTDR Tests

The most accurate and common way of testing loss is by transmission of light from a source and measured by a meter, but one can also imply losses in long fiber cables by backscattered light using an optical time domain reflectometer (OTDR) (Figure 13–5). OTDRs are widely used for testing and troubleshooting outside plant fiber optic cables, for example, in telephone and CATV networks. However, they are not designed to measure the loss of the cable plant or to work on short cables typical of premises cable installations.

Among the uses of OTDRs are measuring the length of fibers and finding faults in fibers, breaks in cables, attenuation of fibers, and losses in splices and connectors. They are also used to optimize splices by monitoring splice loss. One of their biggest advantages is they produce a picture (called a trace or signature) of the cable being tested. Although OTDRs are unquestionably useful for all these tasks, they have limited applications for short cables such as those used in LANs and security systems. They also have error mechanisms that are potentially large, troublesome, and not widely understood.

The OTDR works differently than the time domain reflectometer (TDR) used with copper wire. The TDR measures length and shows events characterized by impedance mismatches, which is important in high-speed copper networks. The OTDR uses the lost light scattered in the fiber that is directed back to the source for its operation (Figure 13–6). It couples a pulse from a high-powered laser source into the fiber through a directional coupler. As the pulse of light passes through the fiber, a small fraction of the light is scattered back toward the source. As it returns to the OTDR, it is directed by the coupler to a very sensitive receiver. The OTDR display

**Figure 13–5.** OTDRs are now small and lightweight but extremely powerful. (photo courtesy of Photon Kinetics)

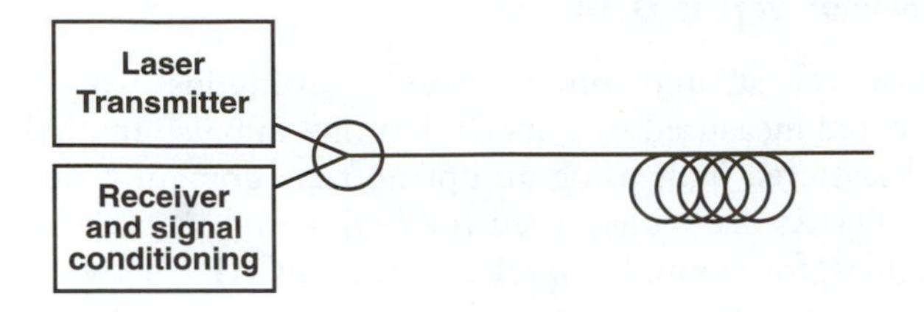


**Figure 13–6.** The OTDR block diagram.

(Figure 13-7) shows the intensity of the returned signal in dB as a function of time, converted into distance using the average velocity of light in the glass fiber.

To understand how the OTDR allows measurement, consider what happens to the light pulse it transmits. As it goes down the fiber, the pulse actually "fills" the core of the fiber with light for a distance equal to the pulse width transmitted by the OTDR. In a typical fiber, each nanosecond of pulse width equals about 8 inches (200 mm). Throughout that pulse, light is being scattered (Figure 13–8), so the longer the pulse width in time, the greater the pulse length in the fiber and the greater will be the amount of backscattered light, in direct proportion to the pulse width. The intensity of the pulse is diminished by the attenuation of the fiber as it proceeds down the fiber, a portion of the pulse's power is scattered back to the OTDR, and it is again diminished by the attenuation of the fiber as it returns up the fiber to the OTDR. Thus the intensity of the signal seen by the OTDR at any point in time is a function of the position of the light pulse in the fiber.

By looking at the reduction in returned signal over time, one can calculate the attenuation coefficient of the fiber being tested. If the fiber has a splice or connector, the signal will be diminished as the pulse passes it, so the OTDR sees a reduc-

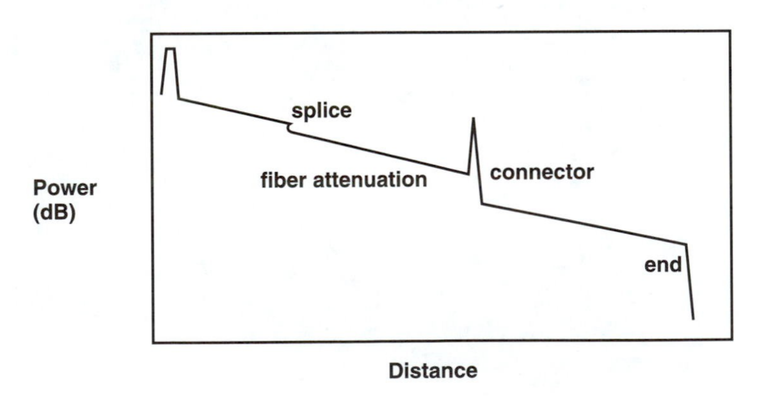


**Figure 13–7.** The OTDR display has information on loss and distance.

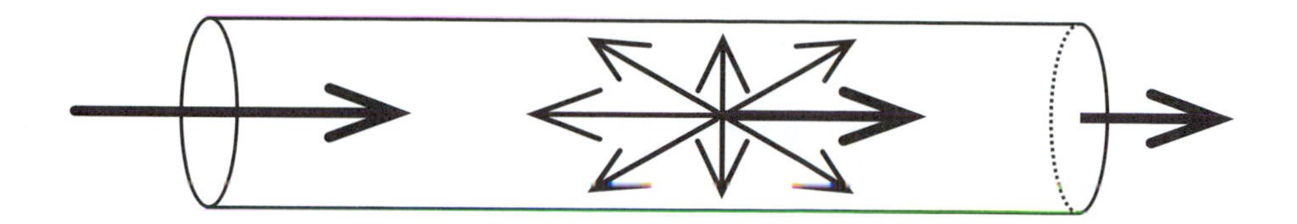

**Figure 13–8.** Scattering of the OTDR pulse sends light back to the instrument for analysis.

tion in power, indicating the light loss of the joined fibers. If the splice or connector reflects light (optical return loss), the OTDR will show the reflection as a spike above the backscattered signal. The OTDR can be calibrated to use this spike to measure optical return loss.

The end of the fiber will show as a deterioration of the backscatter signal into noise, if it is within the dynamic range of the OTDR. If the end of the fiber is cleaved or polished, one will also see a spike above the backscatter trace. This allows one to measure the total length of the fiber being tested.

In order to enhance the signal-to-noise ratio of the received signal, the OTDR sends out many pulses and averages the returned signals. And to get to longer distances, the power in the transmitted pulse is increased by widening the pulse width. The longer pulse width fills a longer distance in the fiber, as noted earlier. This longer pulse width masks all details within the length of the pulse, increasing the minimum distance between features resolvable with the OTDR.

OTDRs send out such a powerful pulse that it overloads the receiver for a certain period of time, when the OTDR cannot see the cable. This is called the OTDR "dead zone." Most OTDRs cannot see within the first 50 to 500 meters, depending on the instrument setup, so a launch cable, sometimes called a "pulse suppressor," is attached to the instrument to allow seeing the beginning of the cable under test. An OTDR should never be used without a launch cable longer than the instrument dead zone.

## OTDR Measurement Uncertainties

With the OTDR, one can measure loss and distance. To use the OTDR effectively, it is necessary to understand its measurement limitations. The distance resolution of the OTDR is limited by the transmitted pulse width. As the OTDR sends out its pulse, crosstalk in the coupler inside the instrument and reflections from the first connector will saturate the receiver. The receiver will take some time to recover, causing a nonlinearity in the baseline of the display. It may take 100 to 1,000 meters before the receiver recovers. It is common to use a long fiber cable called a launch cable or pulse suppressor between the OTDR and the cables being tested to allow the receiver to recover completely.

The OTDR also is limited in its ability to resolve two closely spaced features by the pulse width. Long-distance OTDRs may have a minimum resolution of 250 to 500 meters, while short-range OTDRs can resolve features 5 to 10 meters apart. This limitation makes it difficult to find problems inside a building, where distances are short. A visual fault locator is generally used to assist the OTDR in this situation.

When measuring distance, the OTDR has two major sources of error not associated with the instrument itself: the velocity of the light pulse in the fiber and the amount of fiber in the cable. The velocity of the pulse down the fiber is a function of the average index of refraction of the glass. Although this is fairly constant for most fiber types, it can vary by a few percent. When making cable, it is necessary to have some excess fiber in the cable to allow the cable to stretch when pulled without stressing the fiber. This excess fiber is usually 1 to 2 percent. Since the OTDR measures the length of the fiber, not the cable, it is necessary to subtract 1 to 2 percent from the measured length to get the likely cable length. This is very important if one is using the OTDR to find a fault in an installed cable, to keep from looking too far away from the OTDR to find the problem. This variable adds up to 10 to 20 meters per kilometer and, therefore, is not ignorable.

When making loss measurements, two major questions arise with OTDR measurement anomalies: why OTDR measurements differ from those of an optical loss test set, which tests the fiber in the same configuration in which it is used, and why measurements from OTDRs vary so much when measured in opposite directions on the same splice—and why one direction sometimes shows a gain, and not a loss.

In order to understand the problem, it is necessary to consider again how OTDRs work. They send a powerful laser pulse down the fiber, which suffers attenuation as it proceeds. At every point on the fiber, part of the light is scattered back up the fiber. The backscattered light is then attenuated by the fiber again, until it returns to the OTDR and is measured.

Accurate OTDR attenuation measurements depend on having a constant backscatter coefficient. It is commonly assumed that the backscatter coefficient is a constant, and therefore the OTDR can be calibrated to read attenuation. The backscatter coefficient is, in fact, a function of many factors, including the core diameter of the fiber (or mode field diameter in singlemode fiber) and the material composition of the fiber (which determines attenuation).

The first indication of OTDR problems for most users occurs when looking at a splice and seeing a gain at the splice instead of a loss. Common sense tells us that passive fibers and splices cannot create light, so another phenomenon must be at work. In fact, a "gainer" is an indication of the difference of backscatter coefficients in the two fibers being spliced. The backscatter from the second fiber is greater than the first fiber, and the difference is greater than the loss of the splice.

OTDRs are valuable tools if used properly. Their applications are primarily for outside plant cables over long distances, to verify splice loses and to troubleshoot cable problems. They are *not* designed to be used to measure transmission loss, *nor* are they appropriate for most multimode premises applications.

### Visual Fiber Tracers and Fault Locators

Many of the problems in fiber optic networks are related to making proper connections. Since the light used in systems is invisible, one cannot see the system transmitter light. By injecting the light from a visible source, such as an LED or incandescent bulb, one can visually trace the fiber from transmitter to receiver to check continuity and ensure correct orientation. The simple instruments that inject visible light are called visual fiber tracers or fault locators (Figure 13–9).

If a powerful enough visible light, such as a HeNe or visible diode laser, is injected into the fiber, high loss points can be made visible. Most applications center around short cables, such as those used in telco central offices to connect to the fiber optic trunk cables. However, since fiber tracers cover the "dead zone" where OTDRs are not useful, they are complementary to OTDRs in cable troubleshooting. Since the loss in the fiber is quite high at visible wavelengths, on the order of 9 to 15 dB/km, the fiber tracer has a short range, typically 3 to 5 km.

## CHAPTER REVIEW

1. What instrument is used to test attenuation or loss?
2. What instrument will verify splice loss?
3. What are the three primary wavelengths used with fiberoptics?
4. What components make up an end-to-end cable?
5. What two specifications must you consider in choosing a power meter?
6. Which fiber and wavelength are most sensitive to stress?

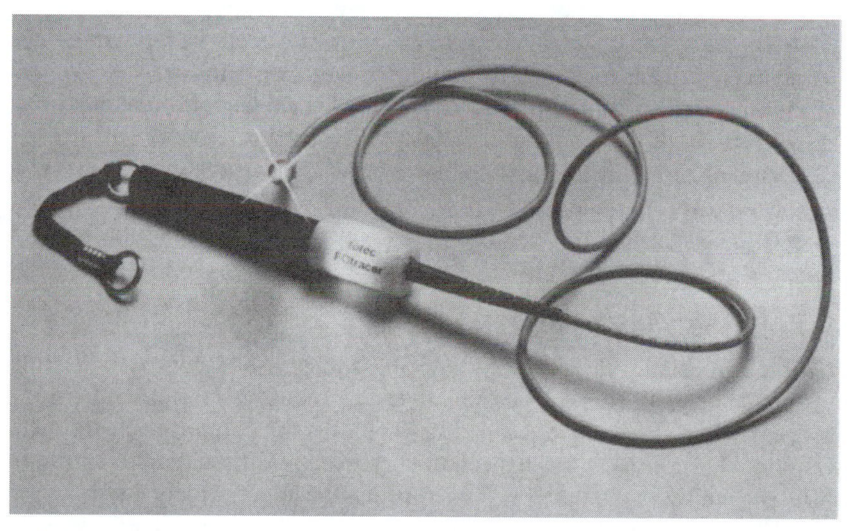

**Figure 13–9.** A simple fiber tracer is a flashlight coupled to the fiber.

7. Which method of calibration is preferred when testing?
8. What does dB stand for, and where is it used?
9. What does the "m" in "dBm" represent?
10. What is the anticipated losses of a connector? a splice?
11. When using a microscope to check a connector end, what would happen if you were to use a magnification that was too high?
12. When viewing through a microscope, which direction allows the best inspection?
13. Where are scratches or defects on the connector end considered to be a problem?
14. What type of instrument uses backscattering?
15. What is the biggest advantage of using an OTDR?
16. What should one add to the OTDR in order for the receiver to recover from saturation from the test pulse?
17. What would you use to assist an OTDR in finding faults too close for the OTDR resolution?
18. Since the OTDR measures the fiber, not the cable, where would you search for a break in a test that showed a break at 15.90 km?
19. What is a "gainer," and what causes it?
20. What is the tool called that injects light down a fiber to check the fiber?

CHAPTER

# 14

# FIBER OPTIC INSTALLATION PRACTICES

The successful installation of fiber optic cabling systems involves pulling or placing cable, preparing the cable for termination or splicing, the termination and splicing itself, and testing or troubleshooting. This chapter will show you how to handle fiber optic cables during installation, splicing, and termination. It will also familiarize you with fiber optic tools and test equipment.

## FIBER OPTIC SAFETY RULES

*Before working with fiber optic cables, familiarize yourself with these safety rules, and follow them to prevent accidents!*

1. Keep all food and beverages out of the work area. If fiber particles are ingested, they can cause internal hemorrhaging.
2. Wear disposable aprons or be very careful to minimize fiber particles on your clothing. Fiber particles on your clothing can later get into food and drinks or be ingested by other means.
3. Always wear safety glasses with side shields. Treat fiber optic splinters the same as you would glass splinters.

4. Never look directly into the end of fiber cables until you are positive that there is no light source at the other end. Use a fiber optic power meter to make certain the fiber is dark. When using an optical tracer or continuity checker, look at the fiber from an angle at least 6 inches away from your eye to determine if the visible light is present.
5. Only work in well-ventilated areas.
6. Contact lens wearers must not handle their lenses until they have thoroughly washed their hands.
7. Do not touch your eyes while working with fiber optic systems until they have been thoroughly washed.
8. Keep all combustible materials safely away from the curing ovens.
9. Put all cut fiber pieces in a safe place.
10. *Thoroughly* clean your work area when you are done.
11. Do not eat, drink, or smoke while working with fiber optic systems.

## TOOLS FOR CABLE PREPARATION AND PULLING

For working with fiber optic cable, you will need a cable slitter to cut and slit the jacket of large cables, Kevlar scissors to cut strength members, a jacket stripping tool for smaller cables, a buffer tube cutter for loose tube cable, and a buffer stripper to remove the plastic coating from the fiber to expose the glass itself. These tools are available individually or in kits from a number of manufacturers (Figure 14–1).

## PREPARING CABLE FOR PULLING

Fiber optic cables are designed to be extremely rugged to protect the fibers, as long as the cable is not overstressed. The most important thing to remember is that cable must be pulled only by the strength members provided in the cables, never by the fibers, and only special cables can be pulled by the jacket. If you are ever in doubt about the proper way to pull a type of cable, contact an applications engineer at the cable company and get specific instructions. Otherwise you may damage the fibers. Damage to a fiber optic cable is never reversible. If a cable is damaged, it must be removed and replaced, which is potentially very expensive.

There are several common types of cable—tight buffer, distribution, breakout, and loose tube cable—that may be installed in a network. Each has aramid fiber strength members in the cable that should be used to attach a swivel pulling eye to connect to the pulling rope or tape. Larger fiber count cables will also have a central strength member that acts as a stiffener to prevent kinking the cable.

### Zip Cord Cable

Many desktop connections will use duplex fiber zip cord fiber optic cable. It looks like AC power cable because it has two simplex tight buffered fibers connected with

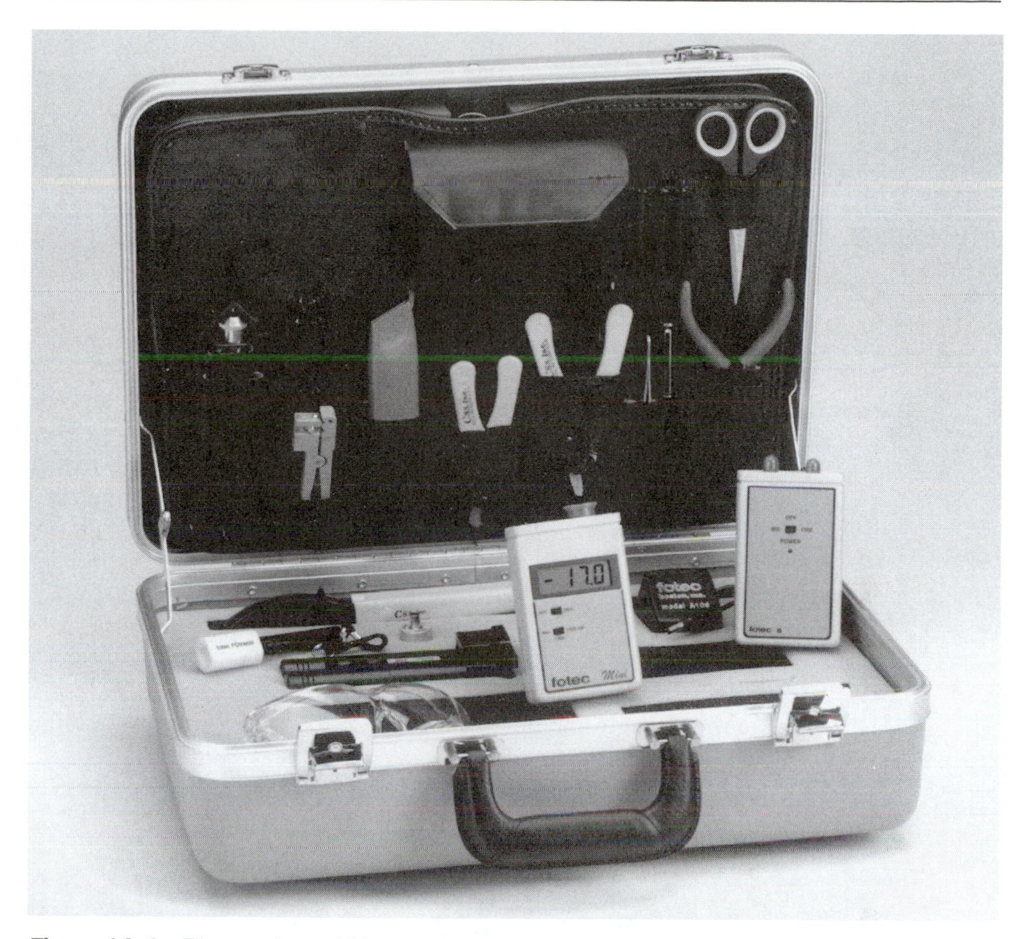


**Figure 14–1.** Fiber optic tool kit.

a central web. It is small and lightweight, and it can be damaged during pulling if done incorrectly.

Like most fiber optic cables, zip cord needs to be pulled by the aramid fiber strength members, not the jacket (Figure 14–2). Since the jacket stretches but the fiber and strength members do not, pulling by the jacket may stretch the jacket, which bunches up the aramid fiber strength members and the optical fiber, causing stress that induces loss and may cause long-term degradation of fiber performance or failure.

Follow these instructions to attach the swivel pulling eye to zip cord cable, as illustrated in Figure 14–3.

1. Split the cable into two single fiber cables.

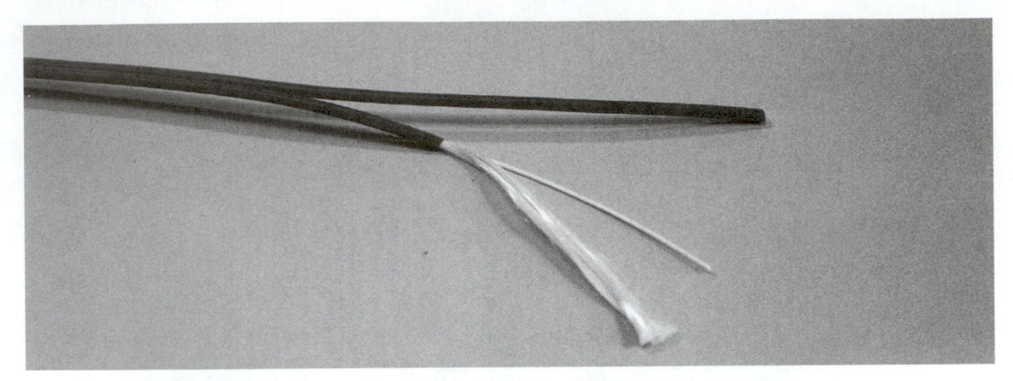

**Figure 14–2.** Zip cord should be pulled by the strength members only.

2. Using the jacket stripper, strip off the jacket exposing the Kevlar (a duPont trade name for aramid fiber) strength members and the buffered fibers. *Do not strip the jacket as wire is stripped—that is, do not cut the jacket then pull off the jacket with the jacket stripper, as it may damage the fiber. Instead, use the jacket stripper to cut almost through the jacket, then use the fingers to rotate the jacket to finish the cut, then gently pull the jacket off the fiber.*
3. Cut off the fibers close to the jacket end.
4. Tie the strength member to one end of the swivel pulling eye (Figure 14-3a).
5. Tape the strength member to the cable jacket to secure the eye. (Figure 14-3b)

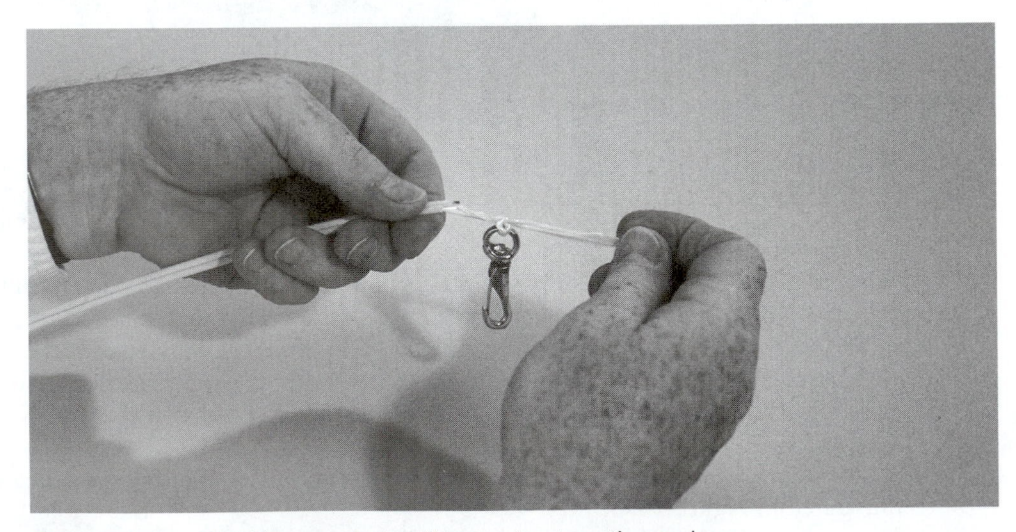

**Figure 14–3a.** Attaching a swivel pulling eye to strength members.

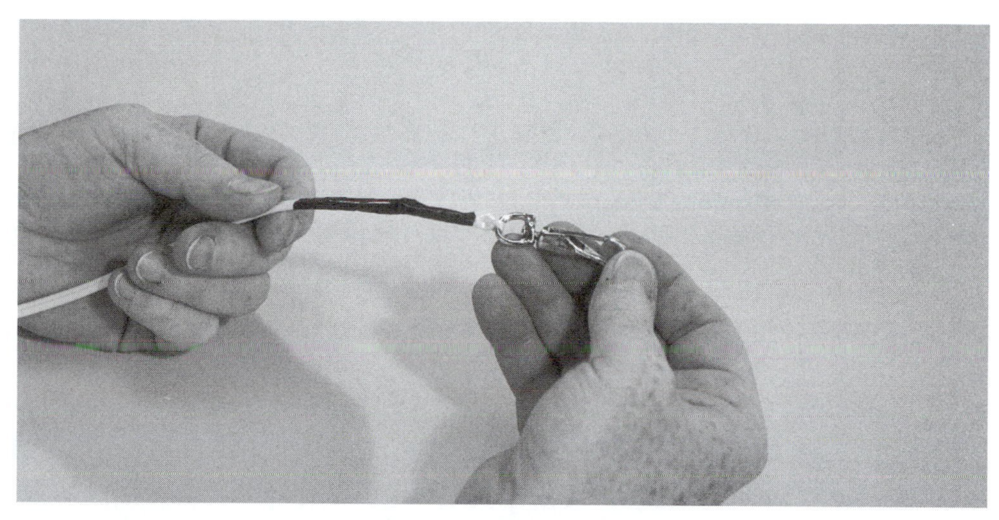

**Figure 14–3b.** Tape strength members to prevent untying.

### Multifiber Cables

Multifiber cables can be of three types—breakout, distribution, or loose tube, depending on the cable construction. Each has a heavy outer jacket for protection and an inner structure to hold and protect the fibers. Working with any of these cables requires removing the heavy outer jackets first.

### Removing Cable Jackets

Preparing the cable for pulling or termination will require removing the jacket of the cable as the first step to expose the fibers. To remove the jacket, use the jacket stripping tool, also called a jacket slitter. Follow these steps, as illustrated in Figure 14–4:

1. Viewing the end of the cable, confirm the outside jacket is round and uniform.
2. Hold the jacket slitter up to the cable jacket and use the knurled nut to set the blade depth to approximately 80 to 90 percent of the thickness of the cable jacket. You do not want to cut through the jacket, as you might damage the cables inside.
3. Make a trial cut of the jacket about 3 to 4 inches back from the end to see if the cutting depth is correct.
4. Place the slitter on the cable and make several turns around the cable. Do not force anything; the spring tension of the tool will cut the jacket gently.
5. Remove the tool.
6. With your thumb under the cable to limit the bending, bend the cable until the jacket snaps.

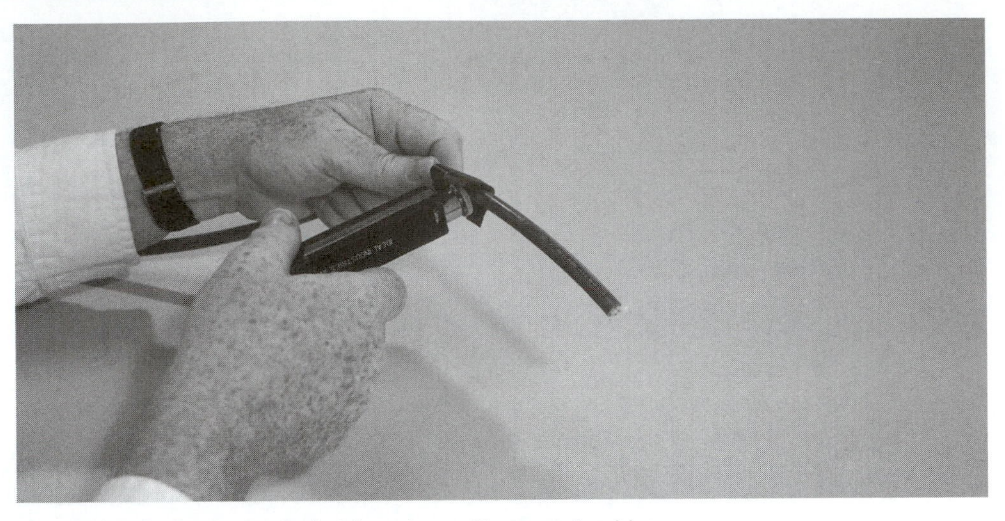

**Figure 14–4.** Removing jacket from large fiber count cables.

7. Turn the cable over and repeat the bend on the opposite side so the jacket is completely snapped.
8. Pull the jacket off the end of the cable.

If the jacket slitter worked correctly, you now have about 4 inches of subcables, a strength member, and a "pull string" or "ripcord" sticking out of the end of the jacket. Use the ripcord to slit the jacket to about 2 feet back from the end (Figure 14–5) by following these steps:

1. Use the needle-nosed pliers to grip the ripcord.
2. Coil the ripcord around the jaws of the pliers.
3. Pull gently back along the cable to slit the jacket.

Use the jacket slitter to cut the jacket just beyond where the jacket was slit by the ripcord and remove the section of jacket. If there is a central strength member, cut it off, leaving only enough to tie off or clamp. Now you have two feet of fibers (in distribution cable), subcables (in breakout cable), or tubes (in loose tube cable) ready to prepare for termination.

### Breakout Cables

Breakout cables consist of a number of single fiber subcables made into a larger cable assembly. Each subcable has a buffered fiber surrounded by Kevlar strength members and a plastic jacket that can be terminated by a standard connector. The subcables are wound around a central strength member that also acts as a bend radius limiter. Some cables can also have two jackets, one for outside plant installations that can be removed to expose a flame-retardant inner jacket for indoor use (Figure 14–6).

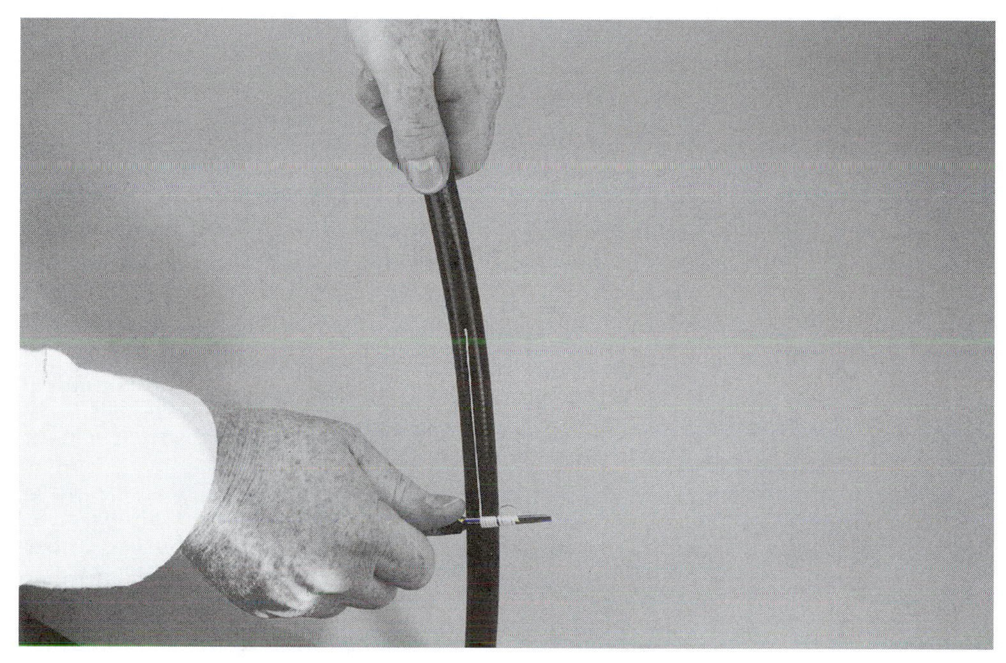

**Figure 14–5.** Always pull the ripcord straight back along the cable to slit the jacket.

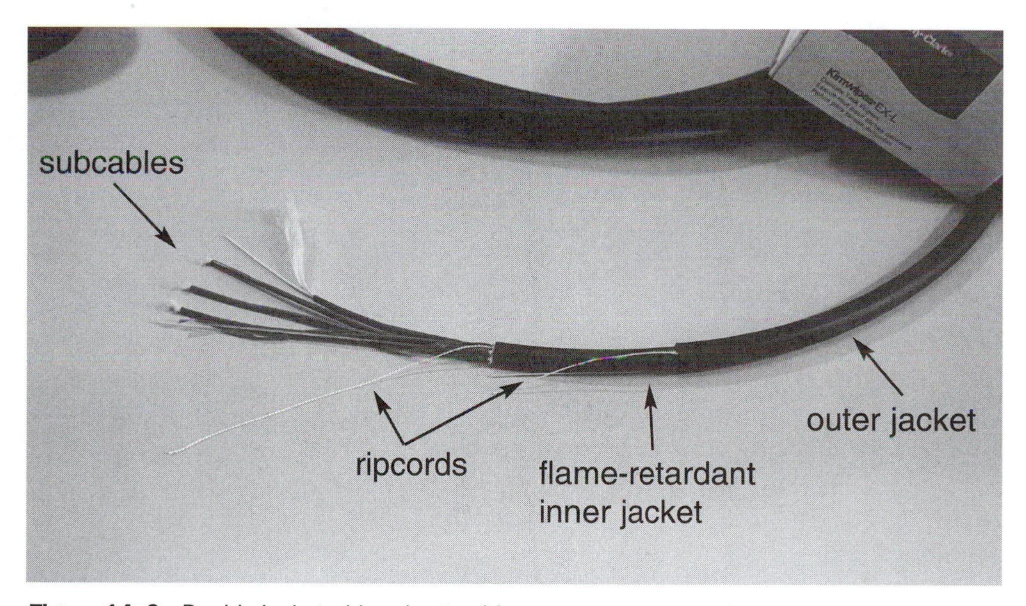


**Figure 14–6.** Double-jacketed breakout cable.

The big advantage of the breakout cable is that it can be brought to a termination point and have the jacket stripped off and individual subcables terminated directly. Then the subcables can be connected to patch panels or terminal equipment with no further hardware. The easier termination at the ends makes breakout cable very cost-effective in many building applications. The disadvantage of the breakout cable is its cost and size. For longer runs, it may not be the best choice.

Pulling breakout cable requires more care because of its more complex construction and larger size. It is usually pulled by stripping the Kevlar strength members from each subcable and cutting off all the fibers, then tying the Kevlar strength members to a pulling eye that is firmly attached to the central strength member. In some cases a jacket gripper is use in conjunction with the pulling eye. Consult the cable manufacturer for special instructions for longer pulls on breakout cable.

### Distribution Cables

Distribution or tightpack cables are designed for use in dry conduit or short riser applications. This cable consists of a bundle of 900-micron buffered fibers with a central stiffener/strength member, a wrapping of Kevlar strength members, and a outer jacket (Figure 14–7). Distribution cables are much smaller and lighter than breakout cables, but the individual buffered fibers require termination inside a patch panel or junction box or sleeving each of the individual fibers in a breakout kit before termination.

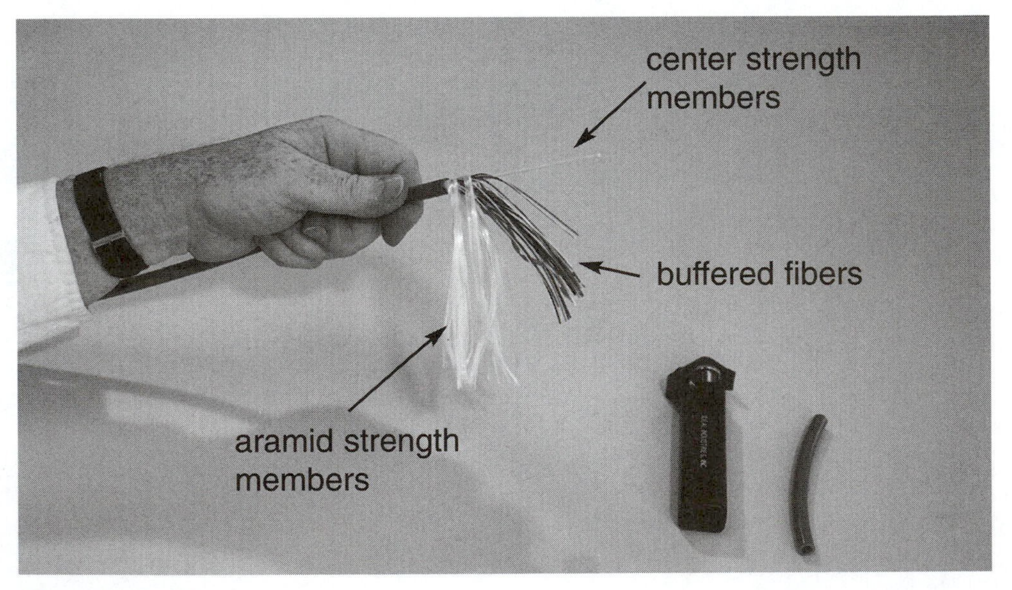


**Figure 14–7.** Distribution cable.

Combinations of the two designs, can be made for some applications. For example, a number of smaller distribution cables can be combined into a breakout cable. At some point, the breakout cable jacket can be stripped and the individual distribution cables pulled to separate locations.

Preparing distribution cable for pulling involves the following:

1. Twist the Kevlar fiber to make it look like yarn and tie a knot in the end to facilitate handling.
2. Use the swivel provided in the toolbox.
3. Tie the swivel to the strength member about 2 inches from the end of the cable with two half-hitches.
4. Pull the Kevlar back over the cable and cut it so that it overlaps the jacket by about 1 inch.
5. Tape the Kevlar over with electrician's tape.
6. Make sure there are no rough edges that can snag on the conduit during the pull. The swivel is now ready to tie to the pulling rope.

### Outside Plant Loose Tube Cable

Outside plant cable is usually loose tube cable with strong dual jackets with aramid fiber strength members or armoring between the jackets. This cable often has a jacket that is strong enough that it can be pulled directly by the jacket using a Kellums grip. Duct cable will be installed in conduit, and aerial cable may have an internal strength member or require being lashed to a messenger wire. Direct burial cable is often armored with a thin layer of metal to prevent rodent damage.

### Armored Cable

Armored cable has a thin metal layer between two jackets for protection against rodent penetration in direct burial installation. The outer jacket and armor are generally thin enough that once a small part is removed, a ripcord can be used to split the armor and outside jacket for easy removal (Figure 14–8). The armor is too hard to cut with a normal cable slitter, so a regular plumbing tubing cutter is used. The tubing cutter blade, which cuts about 1/8 inch deep, is ideal for cutting the outer jacket and armoring without harming the inner jacket and fiber.

This is the procedure for removing the outer jacket and armor:

1. Using the tubing cutter, make a cut about 4 inches in from the end (Figure 14-8a).
2. Keep tightening the cutter just until the shoulder of the cutter reaches the jacket and the cutting blade has penetrated to the full depth. It is not advisable to tighten the cutter any further, as it cannot penetrate further and will merely flatten the cable.
3. Remove the cutter.
4. Flex the cable to finish breaking the outer jacket and armor.

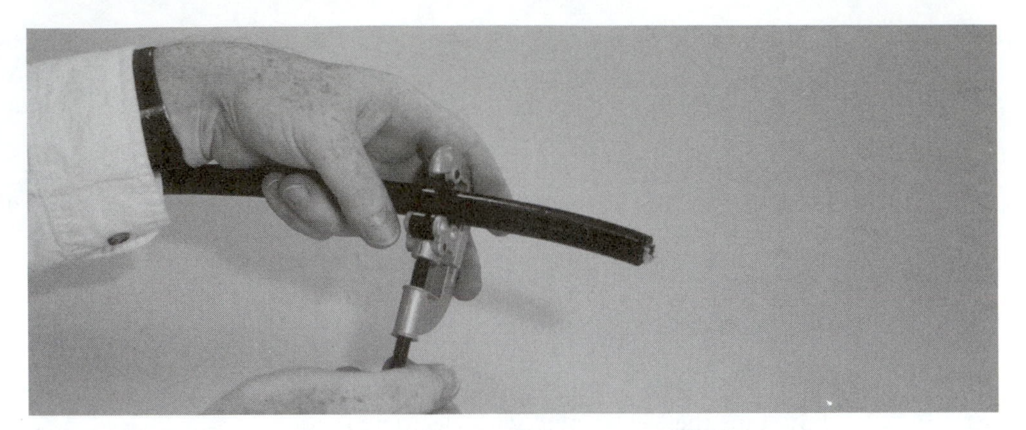

**Figure 14–8a.** Cutting the armor on an armored cable with a tubing cutter.

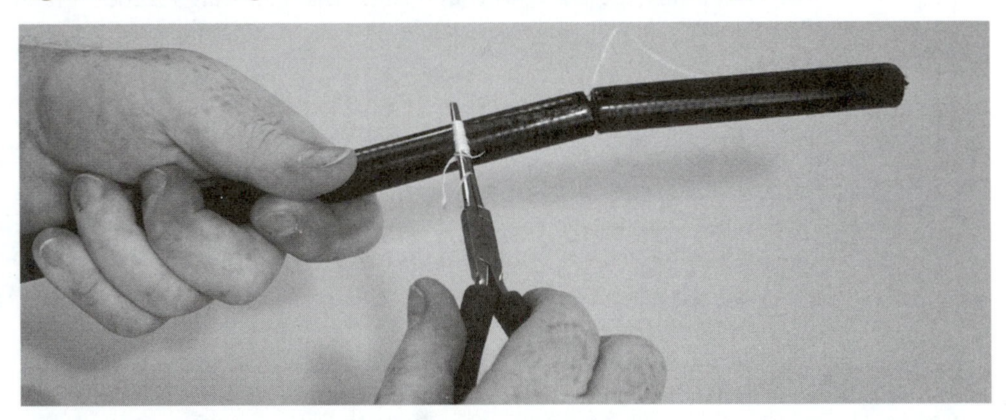

**Figure 14–8b.** Slitting the armored jacket with the ripcord.

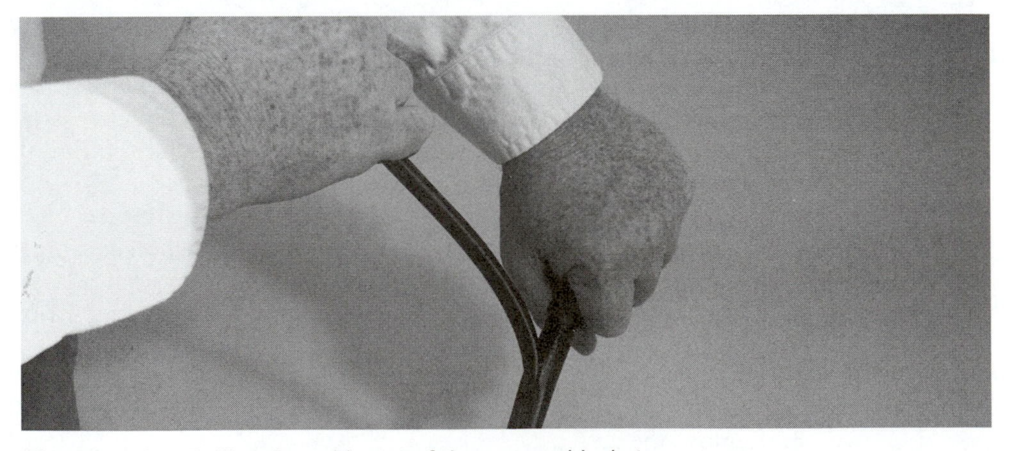

**Figure 14–8c.** Pulling the cable out of the armored jacket.

5. Slide the short section of outer jacket and armor off the end.
6. Use the ripcord to slit the jacket. With the needle-nosed pliers, roll the ripcord around the jaws of the pliers to begin cutting through the jacket and armor.
7. Pull the ripcord back along the jacket of the cable to rip the armor and jacket (Figure 14-8b).
8. Repeat with the other ripcord to finish slitting the armor and jacket.
9. Use the tubing cutter to cut through the jacket and armor just beyond the end of the slit.
10. Pull off the slit armor and jacket segments (Figure 14-8c).

The inner cable can now be handled just like any other cable for termination and splicing.

## PREPARING CABLE FOR TERMINATION AND SPLICING

Before fiber may be terminated or spliced, it must be fully exposed and cleaned. Different types of cables have different means of protecting the fibers, so the methods of exposing the fibers differ.

### Zip Cord or Simplex Cable

Zip cord or simplex cable is generally terminated directly with connectors that are attached to the fiber with an adhesive and crimped to the Kevlar strength members. This is the procedure:

1. Split the zip cord cable into two single fiber cables.
2. Using the jacket stripper, strip off the jacket exposing the Kevlar strength members and the buffered fiber to the proper length for the connector.
3. Cut the Kevlar with special scissors to the length required.
4. Follow directions to install the connector.

### Breakout Cable

To prepare breakout cable for termination, first remove the outer jacket to expose the inner cables for the length needed. Breakout cables have inner cables that are fully as protective as simplex or zip cord, so they can be run for large distances if necessary. Stripping the jacket for long distances is easily accomplished by following the directions given earlier for stripping using a ripcord. If there is a central strength member, it should be cut back as far as the jacket is removed.

The individual subcables should have their jackets stripped, Kevlar strength members cut to the proper length, and the fibers stripped as specified by the manufacturer or the connector being installed, as was done in preparing simplex or zip cord cable.

### Distribution Cable

Distribution cable consists of a central strength member, bundles of buffered fibers, Kevlar fiber strength members wound around the fibers, and an outer jacket. It is very important when cutting the jacket with the cable slitter that you do not cut through the jacket, as the fibers are simply bundled inside and may be nicked by the slitter blade. Prepare the cable as follows:

1. Verify the blade depth of the cable slitter by checking it against the jacket at the end of the cable.
2. Make a trial cut a few inches back from the end to make sure the blade depth is correct.
3. Since the jacket on distribution cable is not tightly bound, you do not have to use the ripcord to slit the jacket, although you may do so if you wish.
4. Cut the jacket about 18 inches back from the end.
5. Break the jacket over your thumb and pull the jacket off the cable.
6. Unwind the Kevlar strength members. They are probably counterwound, so you may want to push back the Kevlar and cut it to the length needed (about 10 to 12 inches to attach a pulling eye, less for tying off at a junction box).
7. Remove the binder tape that holds the bundles of fibers together.
8. Identify the central member and unfold the fibers wrapped around it.
9. For termination, cut the central member off at a length necessary for clamping or tying off.

### Single Jacket Loose Tube Gel-Filled Cable

Loose tube cable for outside plant installation is usually gel-filled to protect the fibers from moisture or water. A single jacket cable cannot be pulled on the jacket, so it is important to separate the strength members if it is being prepared for pulling. It is recommended that you work over a clean work surface with disposable paper, as the gel is messy! This is how such a cable is prepared for termination of splicing:

1. Inspect the jackets for concentricity.
2. Set the cable slitter blade depth to cut both jackets.
3. Take a test cut about 4 inches from the end of the cable.
4. Slide off jacket.
5. Find the ripcord and use it to slit the jacket several feet back (to the length you need for fiber splicing or termination).
6. Use the cable slitter to cut the jacket at the end of the slit and peel off the jacket.
7. You now have a "gooey mess" that can be cleaned with commercially available gel cleaners in wipe form.

8. Peel off the Kevlar strength members, which can be cut to the proper length for attaching to a pulling eye or tying the cable off.
9. Remove the binding tape.
10. Separate the loose tubes that contain the fibers and the central strength member.
11. Cut the central strength member off at the proper length.
12. Clean the tubes where you plan to cut them. The length will be determined by the hardware you are using for splicing or termination.
13. You can cut the buffer tubes by scoring them with the tubing cutter. Let the cutter work on its own—do not force it.
14. Feel where the tube is scored, then place your thumb under it and gently snap the tube.
15. Pull the tube off, exposing the fibers.
16. Wipe the gel off the fibers.

Now you are ready to splice or terminate the fibers. Most loose tube cables will be spliced for outside plant use and are often spliced to premises cable where they enter buildings. Termination requires installing a breakout kit, which protects the fibers for termination. There are numerous types of breakout kits, but all require that the fibers be cleaned carefully to allow fitting the kit.

Make sure you know what hardware you will be using so you can cut the tubes to the proper lengths. Since each tube has several fibers, the tubes are left long enough to extend to splice trays in order to provide protection to the fibers inside the splice closure.

## HANDLING AND STRIPPING THE FIBER

The fiber inside the cable will be protected by a thin plastic coating. A 250-micron buffer coating is applied to the fiber as it is manufactured, and this coating is all that will be found on loose tube cable. Other cable types may have a soft 900-micron coating as further protection.

It is important when preparing the cable to determine the stripability of the fibers. If possible get a sample of the buffered fibers used by the manufacturer to test. The time it takes to strip the fiber will affect the time and cost of the installation job.

When working with cable that has been cut for a length of time, moisture may have caused the end of the fiber to become brittle, making stripping difficult. If the fiber seems brittle, cut off several feet of fiber and try again, until the brittleness is no longer a problem.

There are several types of fiber strippers available, and each works well when used properly. The types that look like wire strippers may need to be held at an angle (Figure 14–9) to prevent stressing the fiber and breaking it, and all types must be kept clean of residue to work properly.

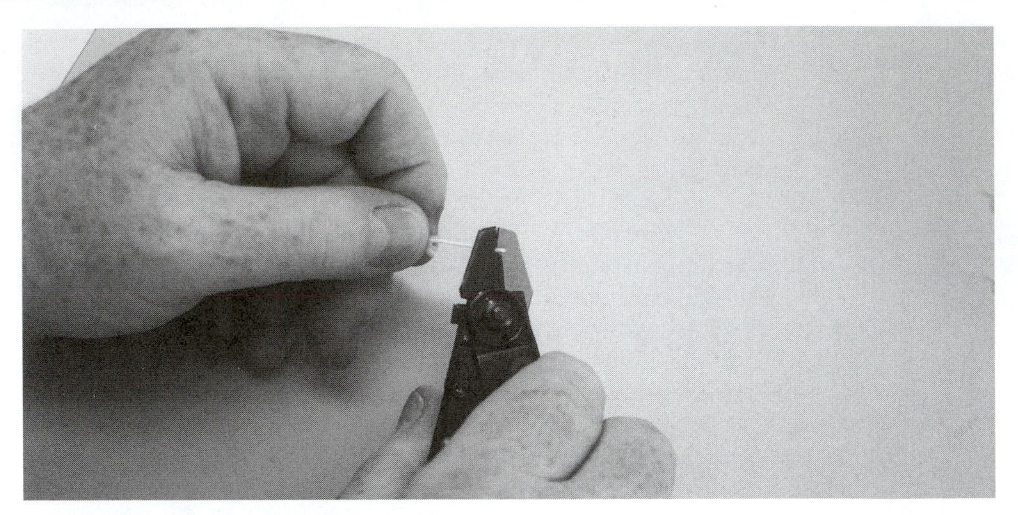

**Figure 14–9.** Stripping the buffer coating off the fiber.

Strip fiber using the following guidelines:

1. Do not use your finger to feel for the fiber ends. You can stick the fiber into your finger and it will usually break off in your finger, producing a painful experience!
2. Always wear safety glasses when working with fibers, to prevent getting fiber pieces in your eye. It is very hard to remove and very painful!
3. To hold the fiber for stripping, hold it between your fingers and wrap it between your fingers in a zig-zag fashion or wrap it around your palm.
4. Do not bend the fiber in a small radius. Fiber is very strong in tension but breaks easily over sharp edges.
5. Use the fiber stripper to carefully remove the buffer coating and expose the fiber.
6. Remove only short lengths of buffer at one time. Longer lengths require greater force to remove and you will risk breaking the fiber. Try removing 1/8 to 3/16 inch (3 to 5 mm) for each cut.

## PULLING CABLE

Pulling fiber optic cable successfully requires observing tension limits and avoiding stress points like corners that may violate bending radius limits. Unless otherwise specified by the manufacturer, fiber optic cable cannot be pulled by the jacket, only by the strength members. This applies to all cable, including simplex and zip cord, but often excluding armored cable which has two jackets and a layer of metal between them. One critical point is to never "jerk" the cable, as it may overstress the cable. If problems are encountered, find the cause of the stress and relieve it.

On long cable runs, it may be preferable to pull the cable from a central point toward both ends. Since it is very important to not put twists in the cable, the cable should be laid out on the ground in a "figure 8" pattern. The figure 8 puts a half-twist in the fiber one way, then takes it out on the other half of the "8," preventing twists.

Indoor runs should be pulled without lubricants if possible, due to the mess they can make in a building. Long runs or difficult pulls should use lubricants. An excellent video on lubricants for pulling is available from American Polywater, Box 53, Stillwater, MN 55082 (612-430-2270 or fax 612-430-3634).

If a pull needs some leverage but not enough to require power pulling equipment, you can use a cable reel, two folding chairs, and a piece of conduit to form a large version of the mandrel puller shown earlier. The cable itself will withstand up to 600 pounds of pulling force. By winding the pulling rope on the cable reel first, the reel will be pulling on the strength member of the cable, not the jacket, and several hundreds of pounds of force can be exerted safely (for both the cable and the installers).

## FIBER OPTIC CONNECTOR TERMINATION

It was not long ago that the proper methods used to terminate fiber optic connectors were tedious and the labor involved was a big concern. However, in the last decade manufacturers have developed new types of cable, connectors, and methods that make fiber termination as easy as copper terminations.

Some of the old methods are still in use today, but progressive installers have quickly accepted many of the new products and procedures. The development of these newer products and techniques is what has led to the accelerated use of fiber in the marketplace.

In this section we will examine the most common method of fiber optic connector termination used in the field, epoxy/polish. Please note that the points examined here are generic in nature and will vary somewhat from manufacturer to manufacturer. We cannot stress enough the importance of following the manufacturers' specific instructions for each type of connector.

We will start with a review of the necessary tools. Each connector type may have a set of tools specific to that connector, but the common tools will work with most epoxy/polish connectors of ST, SC, and FC styles. You should practice termination with a 3-mm jacketed cable, but working with most multifiber cables will be similar, although the strength members may have different uses.

With each type of connector, there are three procedures to follow. Prepare the cable to be terminated, assemble the connector onto the cable, and scribe and polish the assembled connector. The final step is to inspect and test the connector. Polishing is basically the same for each application and each cable type.

Sometimes connectors are directly attached to buffered fiber. The 900-micron buffer of distribution cable is easily terminated, and most connector manufacturers supply strain relief boots for 900-micron terminations. Even 250-micron fiber can be terminated directly, but extreme care must be taken to prevent damage to the fiber. Special breakout kits are available that place small plastic tubes over the fibers to protect them, and although they are tedious to install, they provide much better protection to the fibers.

### Tools

The tools needed for epoxy connector termination are the following: Connector (connector, crimp sleeve, and strain relief boot), crimp tool, Kevlar scissors, jacket strip tool, buffer stripper, scribe tool, polishing tool, microscope (100X), alcohol, low-lint wipes, glass polishing plate, lapping films (12-, 3-, and 0.3-micron grades), epoxy, and empty syringe with needle.

All tools should be laid out in front of you in an orderly fashion. Check at this time to make sure that you are not missing anything.

### Cable Preparation

The cable needed is: 3-mm jacketed fiber optic cable. Prepare the cable as follows:

1. Open the connector package in front of you and take out the parts. If the area is very dusty, do not let the connector fall into the dust as this will clog the fiber hole. Also the connector should have a dust cap on the ferrule. Do not take it off until you are ready to install the connector on the fiber.
2. On each cable end, place a strain relief boot with the small side first.
3. Next place the crimp sleeve on the cable. It will be used to clamp on the Kevlar strength member of the cable and hold the strain relief boot to the connector after assembly. Your assembly should look like that shown in Figure 14–10.
4. Use the jacket strip tool to strip back the jacket of the cable, exposing the needed length of fiber (about 1 1/2 inches). This also exposes the Kevlar strength member of the cable (Figure 14–11).
5. Using scissors made to cut Kevlar, cut back the Kevlar strength members, leaving about 3/8 inch.

*Note:* Never pull on the fiber directly.

### Fiber Preparation

This part will take a bit of practice, but just proceed step by step. You may want to try it several times before trying to make a termination to get a feel for stripping the buffer coating from the fiber. Steps in preparation are as follows:

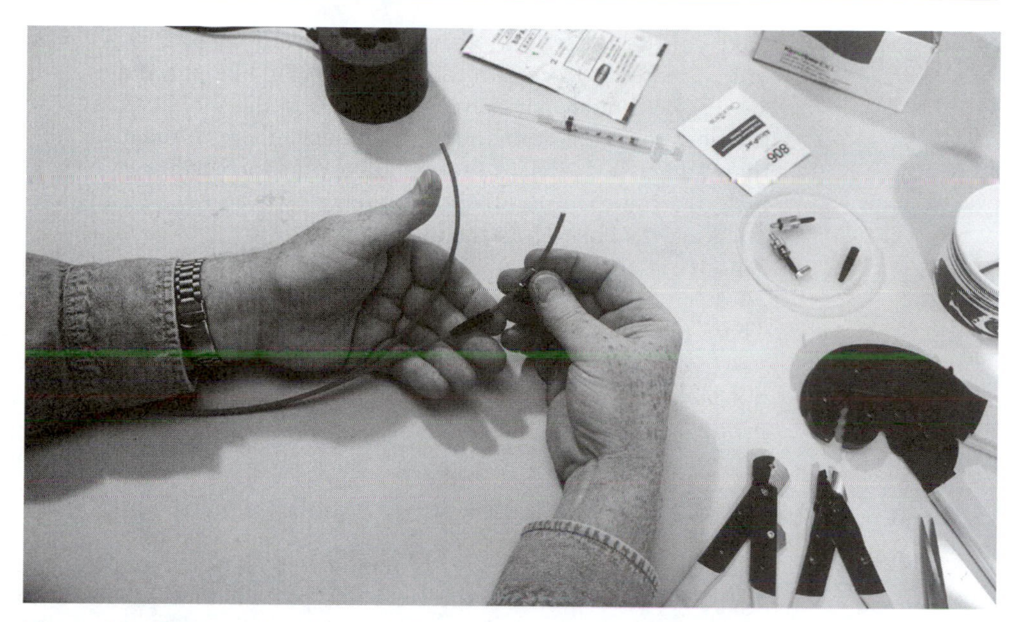

**Figure 14–10.** Place the strain relief boot and crimp sleeve on the cable first.

**Figure 14–11.** Strip off 1 1/2 inches (4 cm) of the cable jacket.

1. Take your buffer strip tool and strip off from .75 to 1 inch (19.25mm) of buffer material from the fiber so it looks like the diagram in Figure 14–12. Be careful that there is no debris in the tool jaws, as it will cause the fiber to break. Some buffer materials adhere to the glass fiber tighter than others. It is advised that you take off short strips of about 1/8 to 1/4 inch at a time. Do not clamp squarely down on the fiber. This will bend and kink the fiber. Hold the tool at a steady angle to the fiber so the fiber is not bent and pull the buffer slowly and steadily down the fiber (Figure 14–13).
2. There may be some debris left on the fiber after stripping. Take a clean lint-free wipe dabbed with a little alcohol and wipe the fiber clean. Do not use rubbing alcohol, as it is mostly water. Use 91 to 99 percent isopropyl alcohol.

### Epoxy Preparation

1. Take the package of epoxy and remove the two-part mix from the package. You will notice that there are two parts with a divider. Remove the divider. Mix the two halves together. If you do not have a tool designed specifically for mixing, such as a roller, the divider may be used. It is extremely important that you completely mix both halves together or the adhesive will not cure 100 percent.
2. Having mixed the epoxy completely, take the empty syringe with needle attached and pull out the plunger. Be careful not to let the plunger roll in any dust.
3. Clip off one corner of the mixed epoxy pack and pour the mixture into the syringe. Do not use the kevlar scissors to cut the epoxy package, as epoxy on these scissors will ruin the cutting edge. Use inexpensive paper scissors instead. When the syringe is full, place the plunger back in the syringe.
4. *Note!* Only place the plunger back in the syringe a very little bit, as it will be full of air. Hold the syringe upside down and let the epoxy run down to the back of the syringe. When the epoxy runs down all the way, you can push the plunger all the way forward, removing the air.

*Note:* Your epoxy has a working time of 30 to 40 minutes.

### Connectorization

1. Remove the dust cap on the connector ferrule.
2. From the back of the connector body, inject the connector with the epoxy. Make sure that the needle of the syringe is inside the connector body as far as it will go. Use light pressure on the plunger as you inject the epoxy until you see a small bead of the adhesive emerge from the fer-

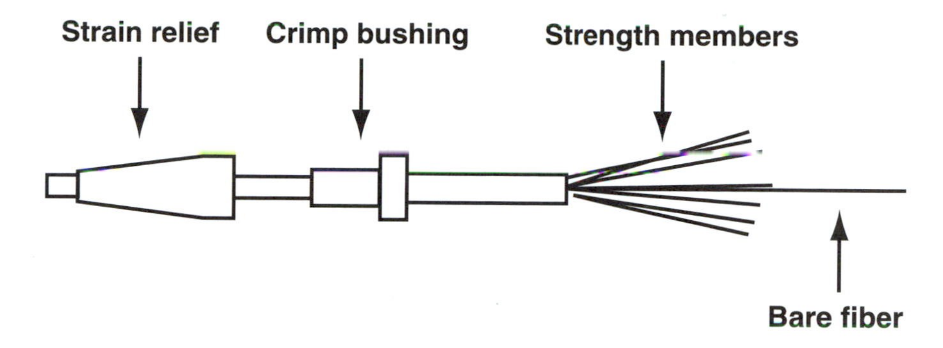


**Figure 14–12.** The cable must have the buffer coating removed from the fiber before a connector can be attached.

rule tip (Figure 14–14). Remove the syringe from the connector halfway and continue to fill the connector until epoxy appears from the end of the connector. Remove the syringe from the connector and pull back on the plunger to prevent any adhesive from coming out of the needle.

3. Insert the stripped fiber through the back of the body of the connector toward the ferrule. Use a twisting motion on the connector to aid the glass fiber in finding the hole in the ferrule. Push the fiber in as far as it will go, rotating the connector on the fiber to evenly distribute the epoxy (Figure 14–15).

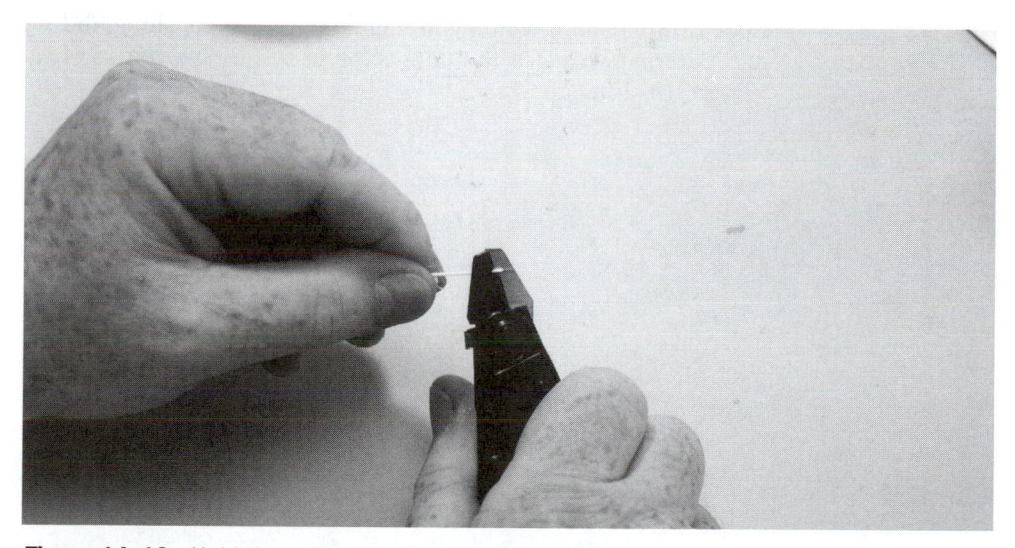

**Figure 14–13.** Hold the stripper so it doesn't bend the fiber and remove short lengths of buffer at a time.

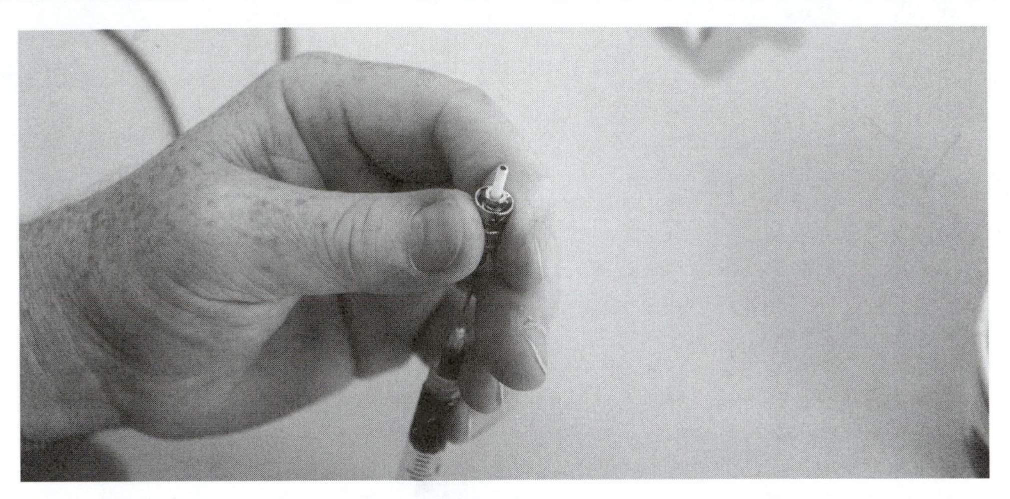

**Figure 14–14.** Inject epoxy into the ferrule until a small bead of epoxy appears on the end of the ferrule.

4. Move the crimp sleeve up over the back of the connector body, capturing the Kevlar, and crimp it to the body using the recommended crimp tool (Figure 14–16).
5. Place the strain relief boot over the back body of the connector.
6. Place a barrel connector over the ferrule of the connector if you are going to leave the connector to cure overnight, to protect the protruding fiber from breakage, as that will ruin the termination. This will protect the fiber from breaking while you handle it before polishing and while curing. Alternately, place the connector in a curing oven for heat curing (Figure 14–17).

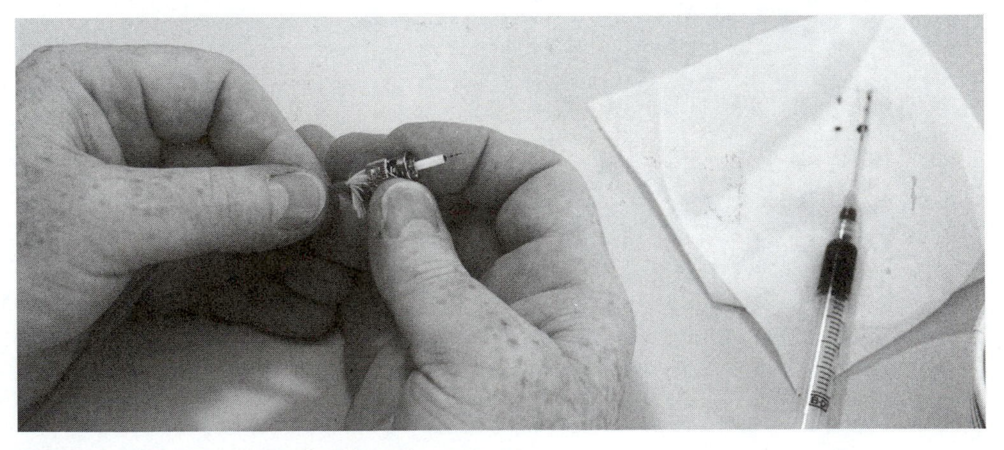

**Figure 14–15.** Insert the fiber into the connector.

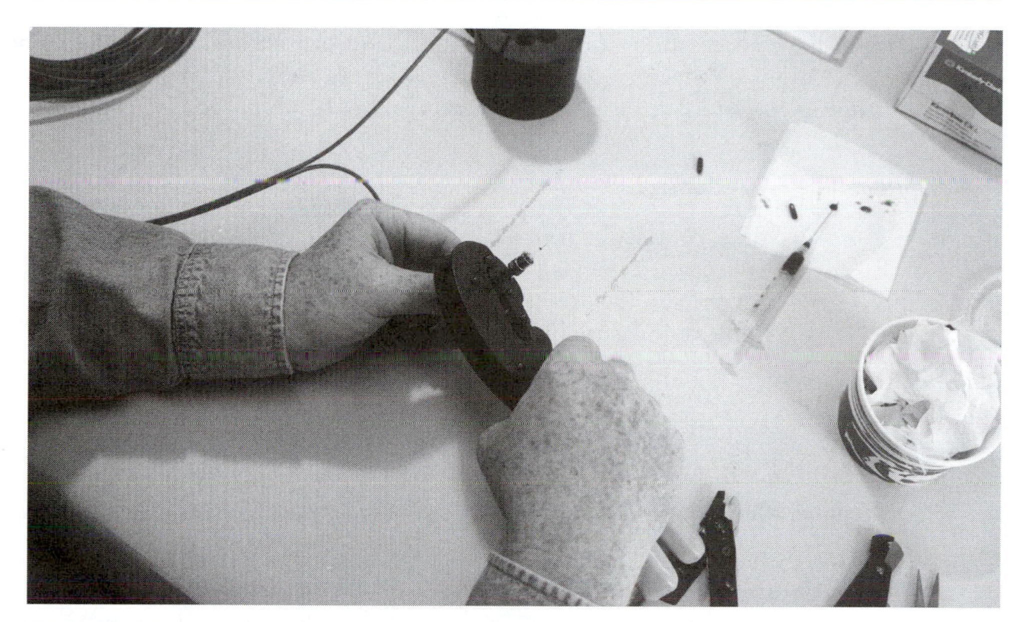

**Figure 14–16.** Crimp the connector to the cable.

**Figure 14–17.** Place the completed assembly into an oven to cure the epoxy.

If you have successfully attached a connector to one end of the cable, do the same for the other connectors. *After approximately 40 minutes, most epoxies will have hardened too much for use, so more epoxy must be prepared.*

*Polishing After Curing*

7. Take your scribe tool and scribe the glass, using light pressure so as not to break off the glass fiber from each connector. Two or three scratches are enough (Figure 14–18).
8. Remove the glass by pulling both up and away.
9. Replace the dust cap at this time.
10. You are now ready to polish the connector ferrule. See the next section on polishing for this procedure.

### Polishing Multimode Connectors

Polishing fiber optic connectors in the field is a simple skill that is easy to learn. However, it takes a "few tricks of the trade" to make it a clean and easy process. The operative word here is *clean*. Try to work in a clean workplace, away from heating vents and dusty processes, if at all possible.

*Note:* Singlemode technology is being used in a lot of data applications. In these situations, concerns over insertion loss are minimized, and there is little con-

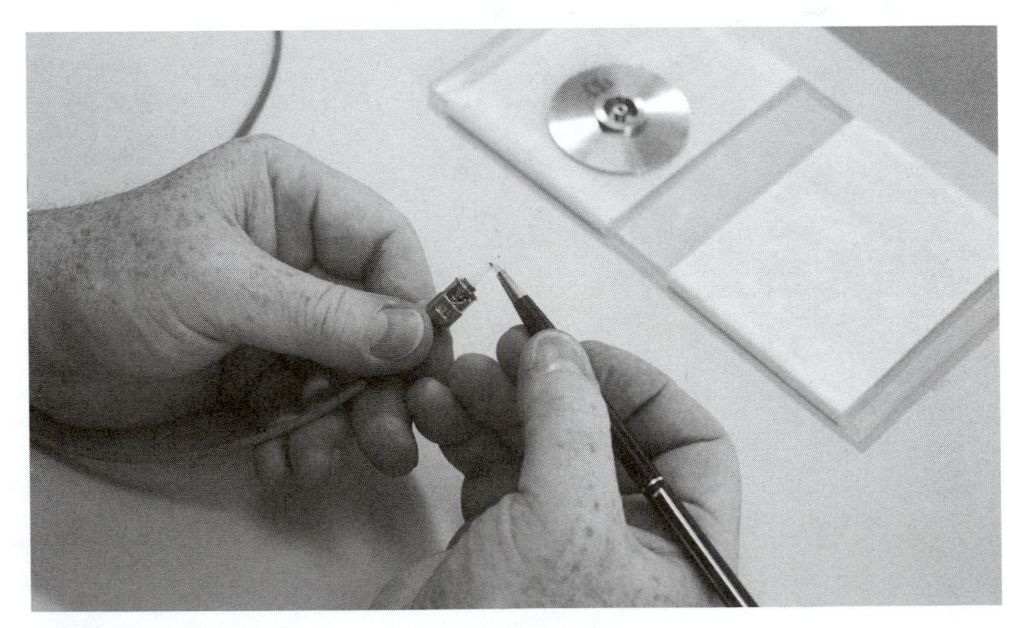

**Figure 14–18.** Scribe the fiber gently—don't break it.

cern over return loss. It is perfectly acceptable to treat a singlemode connector with a multimode polish. When concern over return loss is present, we change our polishing technique to accommodate it.

Lay your tools in front of you in an orderly fashion. You should have a glass polishing plate, lapping films, polishing tool, and lint-free wipes. The polishing procedure is as follows:

1. Place the 3-micron (yellow) and 0.3-micron (white) film on the glass plate. Note that some films are adhesive-backed. You may make a double layer of these to create a softer surface.
2. First "air polish" the connector to remove the glass burr left from cleaving and most of the epoxy bead (Figure 14–19). Take a 12-micron (pink) film and one of the connectors to be polished. Hold the connector upright in one hand and the film with the grit facing down in the other. With very light pressure, polish the face of the connector against the polishing film so as only to take down the glass burr that remains from the scribe step.
3. Now observe the adhesive bead. Polish this down until it is a thin layer but not completely gone.

*Note:* When you are doing an epoxyless connector with a stainless ferrule, polish the fiber down to the metal tip.

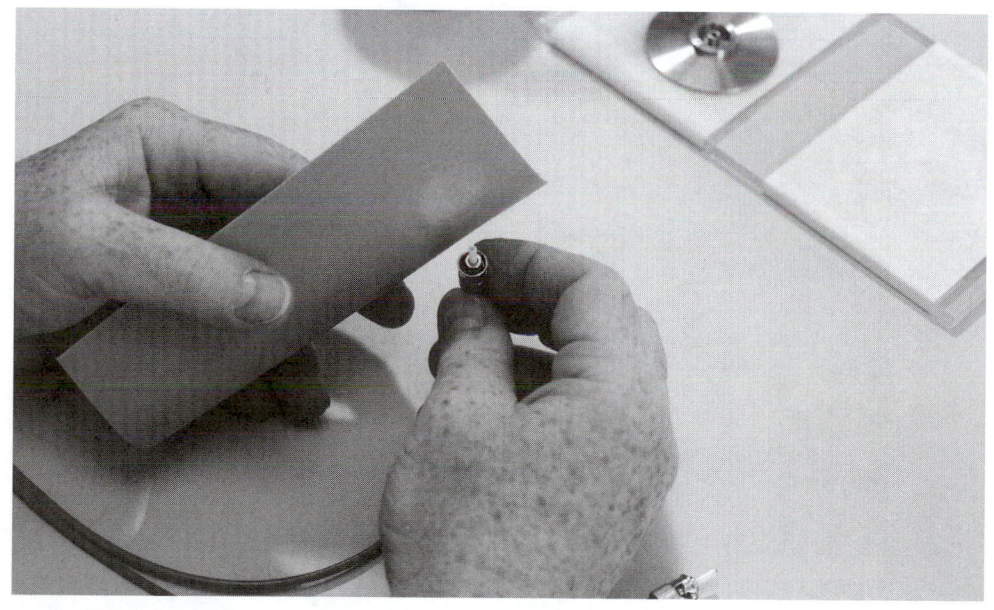

**Figure 14–19.** "Air polish" to remove the fiber stub and most of the epoxy bead.

4. Place your connector in the polishing tool. Lay it down gently on the 3-micron lapping film. Do not slam it down hard, as you could shatter the fiber. Using a figure 8 motion, polish the connector until the adhesive bead is gone (Figure 14–20). If you are polishing a ceramic connector, you will notice that at some point the connector feels slippery on the plate. Stop polishing instantly, since that means you have polished off the protruding glass fiber and epoxy bead and are just rubbing the ceramic ferrule against the polishing film. Now go to the 0.3-micron film. With the same motion, polish with only one or two figure 8s. Your connector is done.

*Note:* With a ceramic connector, you should use almost no pressure at all on the connector as you polish. With a stainless connector, you will want a moderate amount of pressure of about 2 to 3 pounds.

5. Remove the connector from the tool and wipe off both the sides of the ferrule and the face of the ferrule to remove any dust and debris.
6. Observe the connector in your microscope and compare it to the views in Figure 14–21. You should see a clean, scratchless fiber. Backlight the fiber to observe the core; it too should be scratchless. If your microscope permits, view the connector at an angle to assess the fineness of the polish.

**Figure 14–20.** Polish in "Figure 8s".

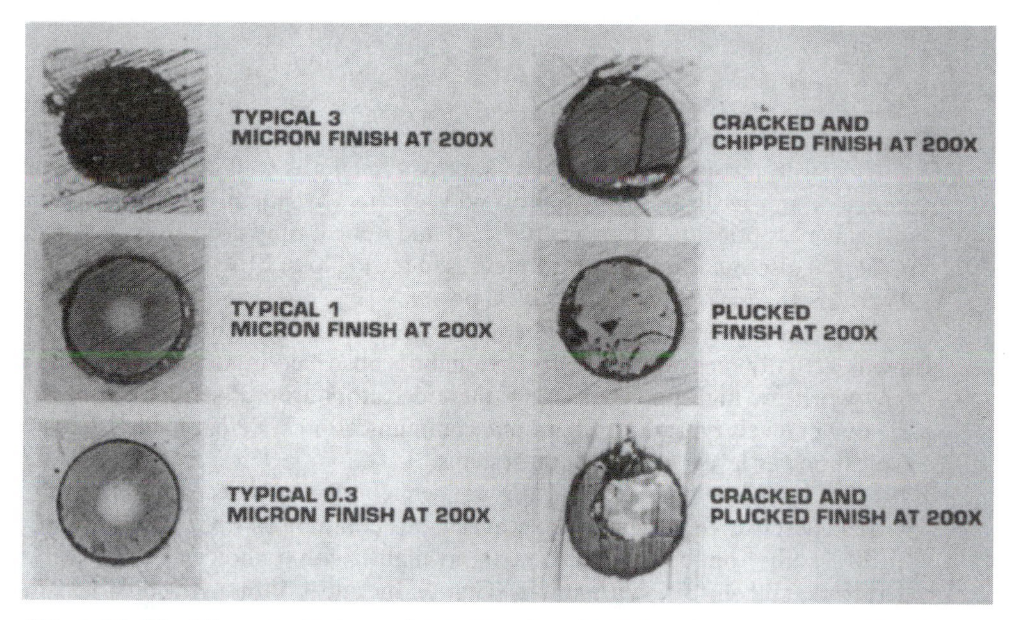


**Figure 14–21.** Microscopic views of connector ferrule and surfaces. *Courtesy Buehler*

## TESTING YOUR CONNECTORS

After you have terminated both ends of a cable, you can test it to see how good your connectors are. You will need a fiber optic power meter, test source, two good cables for reference cables, and two splice bushings to join the connectors.

### Single-ended Loss Test for Jumper Cables

Use the single-ended test method to test your launch and receive jumpers that you will use to test the cables you will terminate later.

A single-ended test (Figure 14–22) uses a matching launch cable on the source to mate with the cable under test. This tests only the connector of the cable being tested that is connected to the launch cable, plus any loss in the cable itself (which is too small to measure in the short cables we use in this exercise).

*Set test power reference*

1. Using alcohol and lint-free pads, clean the ends of all the connector ferrules and replace the dust caps.
2. Attach one of the reference cables to the source. This will be the launch cable.
3. Turn the source and meter on.

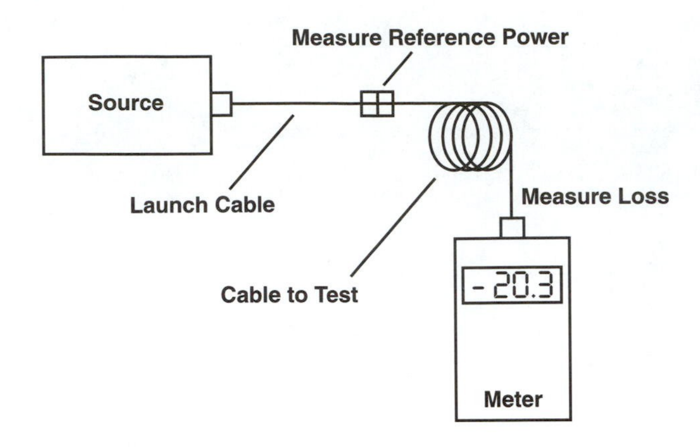


**Figure 14–22.** Single-ended testing of the fiber optic cable.

4. Use the power meter to measure the power out of the launch cable.
5. If the source is adjustable, adjust the source power to a even number like –20 dBm. If it is not adjustable, record the output power of the source. This is your reference power for all loss measurements.
6. Disconnect the launch cable from the power meter.

*Measure single-ended cable loss*

1. Attach a splice bushing (also called a "mating sleeve") to the end of the launch cable.
2. Attach one end of the cable to be tested to the launch cable.
3. Attach the other end of that cable to the power meter.
4. Measure the output of the second cable and record it on the worksheet.
5. Calculate the loss: ___ dBm (test) – ___ dBm (reference) = ___ loss in dB.

*Measure the other direction.* This is called a single-ended test because it measures only the connector attached to the launch cable and any losses in the cable itself. The connector attached to the power meter effectively has no loss, since the detector is large enough to capture all the power from the fiber. By reversing the cable, you can test the other connector. Reverse the cable being tested and test the other direction:

1. Measure the output of the second cable: _____ dBm (3)
2. Calculate the loss: ___ dBm (3) – _____ dBm (1) = ____ dB loss, end B

Is there a difference in the two measurements ? How much ____dB? Remember you are testing the connector on each end separately.

### Double-ended Loss Test for Jumper Cables

Retest your cable assembly with the double-ended method (Figure 14–23), which tests both the connectors on your patchcord, and record the results. Turn the cable around, retest it, and record that data.

A double-ended test uses a matching launch cable on the source to mate with the cable under test and another matching cable on the power meter. This tests both the connector of the cable being tested that is connected to the launch cable, plus any loss in the cable itself (which is too small to measure in the short cables we use in this exercise), plus the loss of the connector on the other end of the cable. *Note: The reference for loss testing is set in the same way as for single-ended testing!*

*Set test power reference (same as single-ended method)*

1. Clean the ends of all the connector ferrules and replace the dust caps.
2. Attach a cable to the source's 850-nm LED. This will be the launch cable.
3. Turn the source and meter on.
4. Use the power meter to measure the power out of the launch cable.
5. Adjust the source power to a nice even number like -20.0 dBm, and record the output. This is your reference power for all loss measurements.
6. Disconnect the launch cable from the power meter.
7. Test another cable for loss using the single-ended method to verify that it is good (typically loss less than 0.5 dB). This will be your receive cable.

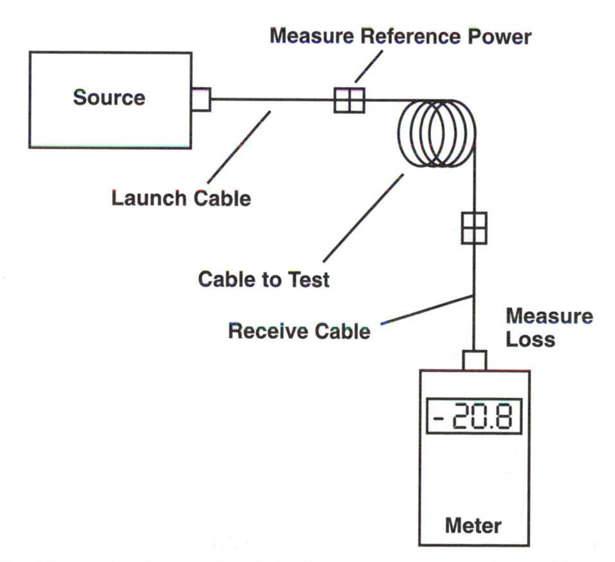


**Figure 14–23.** Double-ended tests check both connectors on the cable under test.

*Measure double-ended cable loss*

1. Attach a splice bushing to the end of the launch cable.
2. Attach one end of the cable to be tested to the launch cable.
3. Attach a splice bushing to the end of the cable being tested.
4. Attach the receive cable to the end of the cable being tested to become the receive cable.
5. Attach the other end of the receive cable to the power meter.
6. Measure the output of the receive cable and record it.
7. Calculate the loss: ___ dBm (test) − _____ dBm (reference) = ____ loss in dB.

*Measure the other direction.* Reverse the cable being tested and test the other direction. You will probably find the readings are different. Each connector you mate with to test a cable has subtle differences that can make the loss different. Since the loss of a connector is the loss mated to a reference connector, it will vary with each reference connector used. Thus connector loss is never an absolute measurement, but only relative to the connector used to test it. Obviously, one should use reference cables with the best possible connections.

## SPLICING

There are two basic categories of splices: mechanical and fusion. Fusion splicing today generally uses an automatic machine that requires considerable training in its operation. If you plan to use a fusion splicer, contact the manufacturer to obtain specific instructions on that machine.

Generally splices are used to connect two fibers permanently. However, there are mechanical splices manufactured to be removable and reusable. We will illustrate splicing using one of those splices.

### Mechanical Splices

Mechanical splices use some alignment mechanism to align two fibers with index matching fluid between the fiber ends. Then a clamp grabs the fibers and/or buffers to hold the fibers in place (Figure 14–24).

There are many styles of mechanical splices. However, they all share some common characteristics. All mechanical splices use an index matching gel or oil to reduce loss and reflections. They are simple to install, requiring only a few basic tools. Typical mechanical splice losses are 0.5 dB or less. The critical element is cleaving the fiber properly.

While mechanical splices require little in terms of specialized tools or fixtures, the splices themselves may be expensive due to the critical nature of aligning the

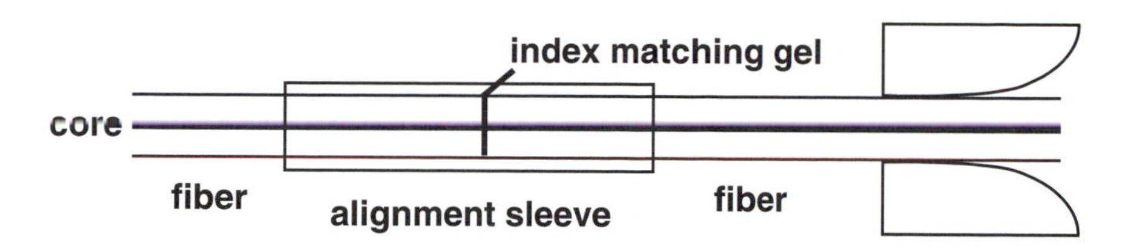


**Figure 14–24.** Mechanical splice.

fiber ends to submicron precision. If you have only a small number of splices to install, mechanical splices may still be less expensive than fusion splicing.

### Splice Installation

Specific installation instructions will vary slightly between manufacturers for their different styles of mechanical and fusion splices. However, every splice follows this same basic procedure:

1. Preparing the cable ends
2. Stripping and cleaving the fiber
3. Aligning and optimizing the splice
4. Fixing the splice to hold the fiber permanently
5. Testing the splice for loss

## EXERCISE

The ideal way to learn splicing is to cut and splice a patchcord, perhaps one made in the termination exercises. Make sure you have tested the cable. Just to be sure, test it again and record the loss.

## PREPARING CABLE

1. Cut the cable in half.
2. Remove approximately 6 inches of the jacket from each end and cut off the Kevlar strength members, leaving the buffered fiber.

(See the cable termination exercise for procedures.)

## STRIPPING AND CLEAVING THE FIBER

The most important factor in getting a low loss splice is getting a good cleave. Cleaving takes practice, so you may want to use some fiber from the cable exercises to practice cleaving before you attempt a splice. Follow the cleaving instructions with your cleaver, and inspect your results with the bare fiber stage of the microscope.

*Do not strip both fibers to be spliced at once. Strip one side, cleave it, insert it into the splice, then strip the other fiber. This is safer and avoids breakage.*

A low loss splice depends on a well-made cleave. Use a microscope and bare fiber stage to view the cleave. A correct cleave should be a clean square cut (Figure 14–25a). Excessive losses will be caused by an angle cleave (Figure 14–25b), a cleave with overhanging material (Figure 14–25c), or a cleave with missing material (Figure 14–25d). Re-strip, scribe, and cleave again if the cleave is not adequate. *Remember, when cleaving always wear protective eyewear so fiber scraps do not get into your eyes.*

## INSERTING THE FIBERS INTO THE SPLICE

1. The cleaved fiber will be inserted into one end of the splice. The cleave length should be such that the end of the fiber is in the middle of the splice. Carefully insert the fiber into the splice; do not force it.
2. Cleave the other fiber in the same manner as before.

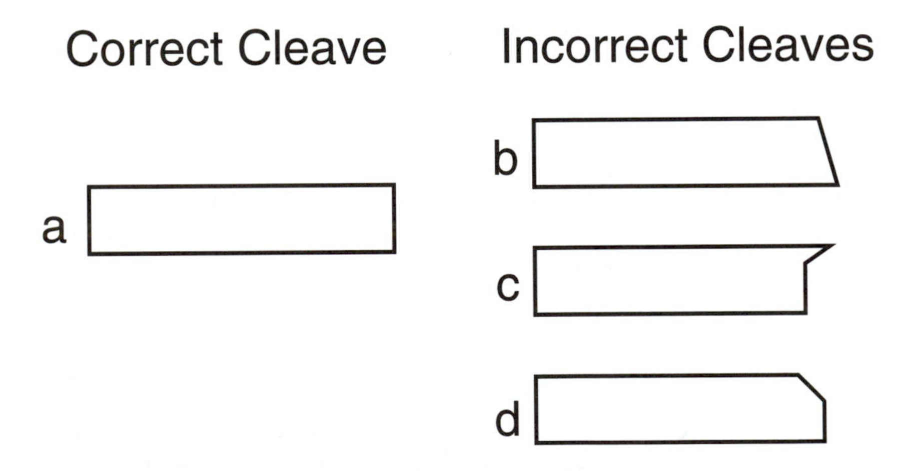


**Figure 14–25.** Proper cleaves are necessary to set low loss splices.

3. Insert the second fiber and gently butt up against the first fiber. Do not force the fibers together, as this will cause them to break.
4. Secure the fibers in place with the mechanism used in the fiber you are using.

## TESTING THE SPLICE

1. Use the meter, source, and one of the reference cables to test the cable that has a splice in it.
2. Record the loss.
3. Compare the loss of the cable before and after adding the splice. Calculate the additional loss of the splice from the loss measured earlier.

## OPTIMIZING THE SPLICE

You can sometimes improve the loss of a mechanical splice by gently withdrawing one of the fibers a slight amount, rotating it slightly, and reinserting it. Try that with your splice if possible.

1. Attach the meter and source.
2. Record the meter reading.
3. Unclamp one fiber.
4. Pull that fiber out by 1 mm (about 1/16 inch).
5. Rotate the fiber 30 degrees and reinsert.
6. Note the power meter reading. Did the loss increase or decrease?
7. Try the process again and note the results.

## CONCLUSION

These are the basic processes necessary for installation of fiber optic cables. If you practice until you can easily perform these tasks, actual installations will be easy. Additional information is available from most manufacturers of fiber optic components, and it is highly recommended you contact them for application notes and especially instructional videos that will familiarize you with the broad spectrum of fiber optic products.

## CHAPTER REVIEW

1. How many safety tips are given?
2. Most cables have a strength member for pulling, and some special cables can be pulled by the jackets, but you should never pull a cable by what?
3. Damage to fiber optic cable caused in pulling can be reversed by sending the cable back to the factory. T or F

4. After you have followed the jacket slitter steps, what do you use to slit the jacket as far as necessary?
5. Which cable is best for short runs inside a building with subcables being connected to patch panels?
6. What type of cable is in a breakout cable?
7. Where are distribution cables used?
8. Why should you be careful with the cable slitter when preparing a distribution cable?
9. What is the common tool that cuts the outside and armor of armored cable?
10. What does the "gooey mess" inside a loose tube cable do?
11. What is required for most loose tube cables for termination?
12. What is the size of the protective buffer around the fiber on a loose tube cable?
13. What makes a fiber brittle?
14. What puts a twist in a cable and then takes it out again for long cable pulls?
15. What are the three basic steps to connector termination?
16. What do you use the scribe tool for?
17. Most of the time singlemode and multimode fibers use the same polishing technique. T or F
18. What should you always do to the ends of the cables before testing?
19. Place in order the basic procedures for every splice—fusion or mechanical:

    ____ Fixing the splice to hold the fiber permanently
    ____ Aligning and optimizing the splice
    ____ Preparing the cable ends
    ____ Testing the splice for loss
    ____ Striping and cleaving the fiber

PART

# 4

# THE BUSINESS OF DATA, VOICE, AND VIDEO CABLING

CHAPTER

# 15

# BUSINESS ISSUES

Even though you may understand the technologies of voice, video, and data communications, there are more things to know before you can actually make money in the communications business. Using your knowledge to make money requires a separate group of skills.

First, you need general business skills. That is, you need to understand finances and where wealth comes from. Although this course does not go into these subjects, they remain critical to your financial success. Here is a brief list of books you should read to educate yourself in business, money, and wealth matters:

*The 22 Immutable Laws of Marketing*, Ries and Trout
*The Richest Man in Babylon*, Clason
*Atlas Shrugged*, Rand
*Think and Grow Rich*, Hill
*Million Dollar Ideas*, Ringer
*Guerilla Marketing*, Levinson

In most ways, communications contracting is the same as traditional electrical contracting or, in fact, any business. Both types of contracting involve the following:

1. Reaching customers and convincing them to use our services
2. Identifying the customers' needs, furnishing them with prices, and entering into transactions
3. Performing services according to the customer's needs
4. Completing the terms of contracts and collecting payment

But communications contracting differs from traditional electrical contracting in how all these things get done. The following sections discuss some of the things you will have to deal with in communications contracting.

## SOURCES OF BUSINESS

One good source of communications business is subcontracting from other electrical contractors. When a local electrical contractor takes on a big project, there usually will be some fiber optic, networking, or other type of communications work that is included, and that may have to be subcontracted. If your company can offer a good price, the electrical contractor may subcontract the work to you. This is often how a company gets started in communications work.

Another source of business is your existing customers. Although the amount of work you can get from them will vary, depending on their businesses, almost all of your commercial and industrial customers use communications systems, as do many residential customers. In addition, these customers typically upgrade their systems every few years.

You may also get referrals from distributors and component dealers. These people often have customers asking them about installers. If your company supplies the vendors with some sales and promotional information, you may get numerous referrals.

## MARKETING

A recent study showed that the average electrical contractor is spending only about four-tenths of one percent of their sales on advertising. No one in communications contracting can get along without spending far more money than that on advertising.

One of the primary methods of advertising is the company brochure. This is perhaps the most general form of advertising in that it can be given to any potential customer.

Another common type of advertising is direct mail. This entails communications contractors sending promotional literature to potential customers in their area. The mailing may be followed with telephone calls to the recipients, and then with another mailing, and perhaps more phone calls as well. Advertising campaigns such as this take time. Results are often slower than you might expect, but these campaigns are usually fairly successful.

In today‘s high tech marketplace, you will be expected to have access to electronic mail (e mail) and to have a company web page. Your web page should have information on your company, personnel, customers (but only with their permission), experience, training and certifications. If you are not familiar with web pages, you can find many consultants who can help you, but today many students, even at high school level, are capable of helping you.

Along with maintaining a trained workforce, advertising is an expense that traditional electrical contractors do not really face. For this reason, mark-up levels must be increased, as additional overhead time and expense will be incurred by the company.

## BECOMING A CONSULTANT

As a supplier and installer of communications systems, your customers will expect you to act more like a consultant than electrical power contractors do. They will expect you to understand electronic systems and to help them find answers to their electronic-type problems.

This will require you (and the other key personnel of your company) to be more of a "techie"—to dress well for customer visits and, in general, behave more like a consultant. You will have to have a broad understanding of electronic systems, so that you can discuss all of your customers' problems with them and find solutions. This is not to say that you have to be an expert in all types of electronic systems, but you must have a broad general knowledge. The idea is not to have every answer (which you cannot), but rather to be able to understand the problem and know how to find the necessary answers.

## EMPLOYEE TRAINING

Getting into communications work requires you to think hard about obtaining properly trained employees. If you wish to use your existing employees, you will have to train them in the systems they will be installing. To do this, you will have to enroll them in some type of training program, whether it be an intensive seminar-type course, a community college program, or correspondence training.

Your other choice is to hire people who are already trained. Eventually, there will be a sizeable pool of fully trained communications installers, but until then, you must try to grab the best ones when they hit the marketplace. This makes the task and the expense of maintaining a workforce quite a bit higher than it is for power wiring, which is reflected in the higher mark-ups that are usually charged for communications work. It may also make extra overhead employees necessary.

## ESTIMATING COSTS

As we go through this chapter, we will be explaining the basic techniques of communications estimating. Many of these steps are common in conventional estimating, but they bear repeating here.

The first step in estimating is to ascertain the overall requirements of the job being estimated. You must get a clear picture in your mind of how this job will flow and, more importantly, where the money will come from, and when. In addition, you must understand the scope of the work for which you are quoting a price—and exactly what will be required of you. These are primary concerns and are the first considerations in any good estimate.

In order to verify that all of these factors have been considered, many estimators use checklists that they review for every project. You should develop your own checklist to use when making estimates.

## VENDORS

When estimating communications work, special attention must be given to the supply channels through which the materials will flow. Conventional electrical estimating requires attention to be paid to this matter, but not nearly to the same extent that is required for communications work. The reason for this is simply that the distribution system for electrical construction materials is highly developed and compact. Finding the materials you need, when you need them, is generally not much of a problem.

The flow of materials for electrical power installations is as follows:

1. The manufacturer buys raw materials from whatever sources are appropriate, and then processes, assembles, and fabricates them until the product is complete.
2. The materials are placed in a warehouse until they are shipped to whatever electrical wholesalers order them.
3. To stay in contact with the hundreds of distributors scattered throughout the country, nearly all manufacturers use a network of manufacturer's rep firms. Each of these representatives works in a certain geographical area, promoting the manufacturer's products and taking orders. The rep firms (called agencies) are paid a percentage of the manufacturer's product sales in their area.
4. The electrical distributors (local supply houses) send out salespeople to call on individual customers in their area. These salespeople are paid on a percentage of their sales and, therefore, have a big incentive to get the customer whatever the customer wants.

Even though there are number of steps in the process we have just reviewed, it is a very efficient process. The end buyer (the electrical contractor) has a convenient channel through which to obtain the materials he or she needs. And should there be any difficulty in obtaining certain materials, the errands and phone calls are done by the wholesale house's salesperson. Thus the contractor can usually deal with the ordering process just once, with only one salesperson.

The communications purchasing process is not so simple. First, two or more links in the distribution chain are sometimes missing. Rare is the communications manufacturer that has a network of knowledgeable reps, and almost never is there a full-service local supply house. Often we think of these extra links in the distribution chain as "middlemen," each of which takes a profit from handling the products—and thus the price of the goods being sold increases. Although there is a certain amount of truth to this idea, it completely ignores the service these people render. The purchase of communications materials is made much more difficult because of the absence of these middlemen.

These gaps in the distribution chain make the processes of estimating and purchasing for communications work considerably more difficult. It is likely that at some future date, more user-friendly product distribution methods will be developed; but in the meantime, the communications estimator will have to deal with the extra difficulties involved in these processes.

Adding to these difficulties is the fact that many communications firms come and go quickly. The manufacture of high-tech products is a rather risky business, and many firms have made great products, only to go out of business a few years later. This makes developing familiarity with any one product and finding replacement parts much more difficult.

In summary, the distribution channels for communications materials are varied, scattered, and sometimes unreliable. This impacts directly on the communications estimator since you cannot properly provide an estimate for something you are not sure you can get. To make the problem somewhat easier to handle, you can use estimating forms that have columns for listing vendors.

Communications estimating requires you to evaluate your sources of supply during the estimating process. Where the material will be coming from is a serious concern for communications work. It is a risk you must attend to if you want to produce reliable estimates.

Frequently, you may have to make allowances for unreliable supply channels. For instance, you may get a price on network interface cards of $105 each. If, however, you are not sure of the reliability of the supplier, it will be necessary to add extra money to the material quote as a safety measure. In this case, you may wish to use a figure of $120, rather than the figure quoted by the supplier. This is where good judgment comes into the process, and where estimating becomes an art as much as it is a science.

## LABOR

The following list contains the essential elements of job site labor. Any good cost estimate must cover these operations. The conventional method of estimating does this conveniently by assigning a single labor unit to each item.

1. Reading the plans
2. Ordering materials
3. Receiving and storing materials
4. Moving the materials to where they are needed
5. Getting the proper tools
6. Making measurements and laying out the work
7. Installing the materials
8. Testing
9. Cleaning up
10. Lost time and breaks

The conventional method of estimating is excellent where circumstances are consistent and predictable. But with communications estimating, there are other factors that come into play. For instance, we can specify a certain labor rate for controller, but there may be other functions required along with the controller, such as programming, software installation, testing, modifications to programs, and so on.

Because of this situation (which is typical to most types of communications equipment), a single labor unit is not sufficient to assign labor for the item. What might be applicable for one installation may be double what is appropriate in the next installation, even though it involves exactly the same type of equipment. The challenge of communications estimating is finding a method of assigning labor in either situation.

In order to solve the problems of charging communications labor, several solutions have been put forward. Although all of them can work well, we recommend that you simply use two lines on the estimate form for each item—one line for a base labor unit (and the material cost), and a second line for "connect labor."

The previous list of the elements of labor contains ten basic functions that we bundle together into one labor unit. Our solution to the high-tech labor problem is to isolate and remove the variable element from the other nine elements of labor, and charge for it separately. By removing operation number 8 (testing) from our labor unit and charging for it separately, we clean up the estimating process. All of our volatility is moved into one separate category, which demands thought before a labor cost is applied; and the rest of the labor, which is consistent and predictable, is covered by the basic labor unit.

As may be obvious, the term *testing* is hardly accurate to describe the types of operations our second labor figure is to account for. Testing would be only one component of these operations. Along with testing would be included any number of technical operations, such as the following;

Installing software
Programming
Configuring hardware
Configuring software
Training users
Running diagnostics
Transferring data
Creating reports

Because of the many operations that can fit into this category, there seems to be no good name for it. So, we are simply using the term *connect labor* to identify this category.

Connect labor is where all of the volatility of communications estimating goes. Connect labor should be calculated and charged separately from normal labor, which we call "installation labor." Installation labor is the labor required to procure the material, move it to the installation area, tool up, and mount the equipment in its place. In short, all installation labor includes the normal elements of labor, but not the connect labor.

## HANDLING SUBCONTRACTS

Communications installations frequently require the use of specialty subcontractors. The reason for this is that there are so many new specialties, no one installer can be knowledgeable in all of them. Therefore, it is likely that all but the largest firms will encounter many situations where there is no one in the company who is knowledgeable about a certain type of application. In these circumstances, you will normally have to call the manufacturer of the system. Manufacturers can refer you to a consultant or dealer in your area who can do the type of work you need done.

## THE TAKEOFF

The process of taking off communications systems is essentially the same as the process used for conventional estimating. By taking off, we mean the process of taking information off of a set of plans and/or specifications, and transferring it to estimate sheets. This requires interpreting of graphic symbols on the plans and transferring them into words and numbers that can be processed.

Briefly, the rules that apply to the takeoff process are as follow:

1. Review the symbol list. This is especially important for communications work. Communications systems are not standardized and, therefore, vary widely. Make sure you know what the symbols you are looking at represent. This is fundamental.
2. Review the specifications. Obviously it is necessary to read a project's specifications, but it is also important to review the specifications before you begin your takeoff. Doing this may alert you to small details on the plans that you might otherwise overlook.
3. Mark all items that have been counted. Again, this is obvious, but many people do this rather poorly. The object is to clearly and distinctly mark every item that has been counted. This must be done is such a way that you can instantly ascertain what has been counted. Color every counted item completely. Do not just put a check mark next to it—color it in so fully that there will never be any room for question.
4. Always take off the most expensive items first. By taking the most expensive items off first, you are ensuring that you will be looking through the plans a number of times before you are done with them. Often you will find stray items that you missed your first time through. You want as many chances as possible to find all of the costly items. This way, if you do make a mistake, it will be a less expensive one.
5. Obtain quantities from other quantities whenever possible. For example, when you take off conduit, you do not try to count every strap that will be needed. Instead, simply calculate how many feet of pipe will be required and then include one strap for every 7 to 10 feet of pipe. We call

this obtaining a quantity from a quantity. Do this whenever you can; it will save you good deal of time.

6. Do not rush. Cost estimating, by its very nature, is a slow, difficult process. In order to do a good estimate, you must do a careful, efficient takeoff. Do not waste any time, but definitely do not work so quickly that you miss things.
7. Maintain a good atmosphere. When performing estimates, it is very important to remain free of interruptions, and to work in a good environment. Spending hours counting funny symbols on large, crowded sheets of paper is not particularly easy; make it as easy on yourself as you can.
8. Develop mental pictures of the project. As you take off a project, picture yourself in the rooms, looking at the items you are taking off. Picture the item in its place, its surroundings, the things around it, and how it connects to other items. If you get in the habit of doing this, you will greatly increase your skill.

## CHARGING TRAINING

Many communications installations require you to teach the owners or their representatives to use the system. This is much more common in high-tech work than in power wiring. After all, you do not have to spend time teaching building owners how to use conventional electrical items, such as light switches, but you will certainly have to spend time teaching them how to use a sound system. Not only that, but you may have to supply operating instructions, teach a number of different people, and answer numerous questions over the phone long after the project is completed.

The question here is whether you include training charges in connect labor, or include these costs as a separate job expense. This decision is essentially up to the discretion of the estimator. It is, however, usually best to charge general training to job expenses, and to include incidental training in connect labor.

If your background is in the electrical construction industry, make sure that you accept these expenses as an integral part of your projects. Do not avoid them. The people who are buying your systems need training, and they have a right to expect it. Include these costs in your estimates, and choose your most patient workers to do the training.

## OVERHEAD

What percentage of overhead to assign to any type of electrical work can be a hotly debated subject. Everyone seems to have his or her own opinion. Whatever percentage of overhead you charge, consider raising it a bit for communications projects. As we have already said, the purchasing process is far more difficult for communications work than it is for other, more established types of work. In addition, there are a number of other factors that are more difficult in communications work than they are in more traditional types of work.

Almost every factor we can identify argues for including more overhead charges for communications work—not necessarily a lot more, but certainly something more. When you contract to do communications installations, you are agreeing to go through uncharted, or at least partially uncharted, waters. This involves greater risk. And if you do encounter additional risk, it is only sensible to make sure you cover the associated costs. You do this by charging a little more for overhead and/or profit.

## MODIFYING LABOR UNITS

Labor units are based upon the following conditions:

1. An average worker
2. A maximum working height of twelve feet
3. A normal availability of workers
4. A reasonably accessible work area
5. Proper tools and equipment
6. A building not exceeding three stories
7. Normal weather conditions

Any set of labor units must be tempered to the project to which it is applied. It is a starting point, not the final word. Unusually poor working conditions typically require an increase of 20 to 30 percent. Some very difficult installations may require even more. Especially good working conditions, or especially good workers, may allow discounts to the labor units of 10 to 20 percent, and possibly more in some circumstances.

For a list of labor units, see Appendix A.

## BIDDING

There are many types of bids that you will be asked to furnish in the communications market. Most of them will be lump-sum bids or unit-priced bids. But there is another type of bid that you will encounter, the RFP.

RFP stands for Request for Proposal. In many specialties (and in the data networking business in particular), RFP is almost equivalent to bid documents, except not as detailed. When someone wants a bid from you, he or she will send you an RFP. This document will give you the general details of the project, and you will be asked to furnish a complete design, schedule, and price.

Completing the RFP process is very similar to performing a design/build proposal. When you prepare RFPs, take care that you will not be doing the design work for the customer (for free), only to have another party do the installation according to your design.

## OPENING A NEW COMPANY

Step number one is to decide which of the specialty markets you wish to enter. Knowing that the profits in such markets are good is one thing, but identifying a specific gap between supply and demand and actually deciding to jump in are quite different. Here are the factors that you need to consider before diving in:

1. The amount of work available—current and future. You want to choose a specialty that is growing, and will continue to grow for a long time. As long as growth is occurring, supply and demand should remain in your favor. You can determine this by doing a bit of research, and finding projections for the various trades. In order to get the projection numbers you want, you can:
   a. Call or visit a reference librarian and ask the librarian to perform a search of various databases.
   b. Find trade journals in that specialty, and search for answers.
   c. Call trade organizations in that specialty.
   d. Post requests for information on special computer bulletin boards. You want to find businesses that are already busy, and will continue to grow. Do not choose a business that is almost nonexistent now but will be huge later—you need something now.
2. Skills and training. Step number two is to find out how much difficulty you will have in training yourself and your workers to perform this work. Your first level of training is for yourself, so that you will be at least as proficient as any of your employees. Your second level of training is for your employees. Calculate your time and expense to get everyone properly trained.
3. Equipment and tools. Identify the types of equipment and tools you will need to get started in the new trade. Specify prices for each item, and come up with a total; it does not have to be as accurate as a bid, but it should be reasonably close to reality.
   Remember that as you start up, you may be able to rent expensive pieces of equipment and tools rather than purchasing them. For instance, if you wish to get started in data cabling, you do not need to buy all of the expensive testers; just verify that they are locally available on a rental basis, and eliminate them from your "must purchase" list.
4. Supply channels. Next, you must identify your supply channels. We have been spoiled with electrical materials; there is no difficulty at all in finding almost any electrical item we might desire. This is not the case in the high-tech specialties. It is not terribly difficult to find what you want in

these trades, but it is certainly not as easy as running by the supply house on the way to the job.

Make a list of the suppliers that you will need to do business with in order to perform your specialty work. Pay attention to delivery times. For instance, if you have to buy mail-order, you will have to order materials long in advance of the installation time. On the other hand, items that can be bought from a local supplier can be ordered just a day or two in advance.

5. A list of local customers. Your next step is to prepare a list of customers in your area. Invest some effort in this, and come up with a good list.
6. Marketing techniques. Define your techniques. How can you best reach the people on your customer list? Consider options such as direct mail, sales calls, bidding, and so on. Also try to determine what your costs are likely to be for a serious marketing campaign.
7. Analysis of the data. Now you need to analyze the data you have developed. Rate each specialty you are considering in each of the previous six factors. At the end of your analysis, you should be able arrive at an intelligent decision about which specialty or specialties you will enter.

### Company Structure

Once you have made the first essential decisions, you need to decide how your new company will be structured.

It is certainly possible to run the new types of work within your existing company, but as a practical matter, it is almost always better to separate the two different types of work. Even if you will still be using the same shop, installers, and tools, you will certainly want to keep separate sets of books. Those contractors who have entered into communications installations have almost all gone to a separate company structure.

For general business and financial reasons, the new company should be incorporated, not a proprietorship or a partnership. Limited partnerships sometimes work, but serious legal expenses are required to establish them.

Now you are faced with the hard questions. Who will be the owner of the company? If more than one person will own it, who will own how much? How much stock will be offered? These questions can be difficult. Get a lawyer to handle them for you. A lawyer's advice in these matters is necessary.

### Infrastructure

Setting up a new company can be a lot of fun. It allows you to build the perfect company from the ground up. You can do in this company everything you have always wanted to do, with no resistance from the status quo.

Your goal will be to set up a system of operating that covers everything that the company needs done. Make a list of all the activities that will be necessary (marketing and sales, estimating, billing, collecting, accounting, purchasing, answering the phones, supervision, and so on). Assign each of these activities to a specific person. Then compare the lists to see how it will all work together, and modify them where necessary. If you can assign all of the duties and provide each employee with a list of responsibilities, your company will be easy to run.

Also identify the computer programs you will use, and what types of computers you will need. Try to get as much of your work as possible off paper and onto disk. If you do this from the very beginning, it should not be especially difficult.

At this time, you will have to make arrangements for your location. Most contractors like to run their spin-off company from their existing shop to avoid new rental and utility expenses. (When starting a new company, it is generally advisable to avoid spending any extra money.) If, however, you have no more room, you will have to rent a separate facility, which will increase your start-up costs significantly.

You must also set up your estimating and job pricing methods. Estimating communications projects is different than estimating power work. Be sure that you know what you are doing before you start throwing prices around. No matter how good your market may be, if you quote bad prices, you will get stung.

### How Long It Takes

This is a lot of work. Starting any type of business should be taken very seriously (in spite of the fact that most of us did not do any research before we got into the electrical business). The goal of the research is to eliminate all possible risks and define the best methods of conducting business before actually spending any money.

The first part of your research and planning should take only a day or two, although gathering your market projections may include many days of waiting. The legal matters may take a few weeks before they are completed by the attorney.

Once you have all of your planning done, you can prepare to conduct commerce. Your first duty is to make sure that you are both able and available to perform the actual installations. Get the necessary tools and/or equipment, and make sure that your people are trained and ready to go. At this point, your shop (or shop area) will need to be ready to go as well. Have the materials and tools laid out in a sensible manner.

Once you are ready and able to conduct business, you can begin to market your services. I strongly suggest that you do not begin marketing until your ability to perform the services is ensured. One of the worst things you can do in new markets is to promise something you cannot deliver. In most cases, this will mean that your people will be all ready to go but will then have to wait for a job to roll in. Tell them in advance that this will happen, and it should cause no hardships.

### Selling

Selling specialty services is a lot different than bidding power projects. You will have to develop a marketing plan and stick with it. (You will, of course, have to modify it as you progress.)

In this new venture, you are more of an entrepreneur than a contractor, so act like one and think like one.

Do not expect instant results with your advertising. Projects like these take time and effort. Sometimes they take a while to pay off.

### First Steps

Prepare yourself for a slow start. New companies take a while to get moving. If your first advertisements do not get you any business, try a different angle and keep going. Do not get scared if your first results are slow in coming. If you have done your research well, your work will soon be rewarded.

Once you begin to get work, be sure to go very slowly. These are new types of projects to both you and your workers. There are certain to be many new, confusing, and difficult situations that will arise. You will need time to deal with these as they appear. If you move slowly, you can resolve these difficulties without creating problems for your customers. If, however, you go too fast, you will not be able react to problems fast enough, and you will disappoint your customers.

Once you get past your first series of projects, you can begin to expand. Even so, do not expand too quickly, as handling a number of projects at the same time will also present obstacles to be overcome. Do not be in a hurry, and in a year or two you can be very profitable.

## SUPERVISING

When you begin to take on projects, it is critical that you know how to supervise the work of your people. Since we are familiar with power wiring, we think of supervision primarily as making sure that the work gets done on time. Making sure the work is done right is a consideration as well, but it is not the biggest thing in our minds. With communications work, however, getting it right is more difficult than getting it done in time.

The problem is that mistakes in communications work (terminations in particular) are difficult to detect. A mistake that could keep the entire system from working might not show up at all until the system is completely installed and turned on. (Think about that for a minute—it is a scary situation.)

So, properly supervising a communications installation means that you must be able to make sure that your work is good. Yes, it must be done on time; but being done on time is meaningless if the work has to be replaced.

Job number one is to ensure that all the terminations are done correctly. This is by far the most important part of supervision. Make sure that your people have all the right parts, that they do not rush, that they have a well-lit work area, and that they have test equipment and use it!

Next, be very sure that your people mark every run of cable and termination well. Spend money on cable markers and numbers and spend time on written cable and termination schedules—do not lose track of which cable is which.

Look over the shoulders of your people on the job to make sure they are doing things right. Terminating communications is fine work; make sure that your people work like jewelers, not like framing carpenters.

## INSPECTION

Inspection is a bit of a wild card in communications installations. Electrical inspectors do not always inspect communication wiring. Nonetheless, take a moment to check with a local electrical inspector before you work in their jurisdiction. And obviously you should be very familiar with the requirements of article 800 of the *NEC*®.

In most cases, the inspector of your communications installation will be the same person who signs your contract. In some cases, it will be a third party. No matter what, make sure that you know who will inspect your work before you give your customer a final price. You must know what the inspector will expect of you, and what he or she will be looking for. Be especially careful of third-party inspectors, since they are getting paid to find your mistakes.

The bottom line in installation quality is getting good signal strength and quality from one end of the network to the other. But be careful of other details that may be noticed by the inspector. Among other things, many inspectors will give a lot of attention to proper cable marking, mechanical protection, and workmanship. Pay attention to any detail that the inspector is likely to examine.

## CHAPTER REVIEW

1. Briefly list the four contracting steps that are similar in communications and electrical contracting.
2. How often do most customers upgrade their communications service?
3. What are the two suggested ways of marketing?
4. As a consultant of communications systems, what else should you be knowledgeable of?
5. In estimating a job, what must the estimator have a clear picture of?
6. What makes estimating for the electrical power installation contractor easier than for the communications installer?
7. What two lines should be used for labor in an estimate?
8. Of the ten basic functions bundled into one labor unit, which one was suggested to be charged separately and why?
9. What is another name for base labor?
10. What name is given to the process of taking information from a set of plans and/or specifications?
11. Briefly list the eight rules for a takeoff process.
12. What is an RFP?
13. In starting a new business, where should you invest some effort?
14. Who should you check with for inspections?

# GLOSSARY

λ **Lambda** = Wavelength of light = *c/f*.

**μm** Micron; one - one millionth of a meter.

**10BASE-2** 10 MB/s Base Band Transmission, 185 meters max. segment length on Thinnet Cable.

**10BASE-5** 10 MB/s Base Band Transmission, 500 meters max. segment length on Thicknet Cable.

**10BASE-F** 10 MB/s Base Band Transmission, 2,000 meters max. segment length on fiber-optic cable.

**10BASE-T** 10 MB/s Base Band Transmission, 100 meters max. segment length on Cat 3 or better twisted pair cable.

**100BASE-T** 100 MB/s Base Band Transmission, 100 meters max. segment length on Cat 5, twisted-pair cable, also referred to as Fast Ethernet.

**1000BASE-LX** Gigabit Ethernet on optical fiber at 1,300 nm.

**1000BASE-SX** Gigabit Ethernet on optical fiber at 850 nm.

**1000BASE-T** Gigabit Ethernet on Cat 5 UTP.

**110 block** Punchdown block from AT&T 110 Cross Connect System.

**66 block** Punchdown block from AT&T 66 Cross Connect System.

**Absorption** That portion of fiber optic attenuation resulting from conversion of optical power to heat.

**ACR** Attenuation to crosstalk ratio; a measure of how much more signal than noise exists in the link, by comparing the attenuated signal from one pair at the receiver to the crosstalk induced in the same pair.

**Adapters** A type of balun that physically allows one connector to mate to another.

**ADO** Auxiliary disconnect outlet.

**ADSL** Asynchronous digital subscriber line.

**ALPETH** Aluminum shield polyethylene (jacket cable).

**American wire gage (AWG)** An American system of defining the size of copper wire.

**Analog** Electrical signals that are continuously varying in format, as opposed to being digitally encoded.

**ANSI** American National Standards Institute; oversees voluntary standards in the United States.

**APD** Avalanche photodiode.

**APS** Automatic protective switching.

**ASTM** American Society for Testing and Materials.

**AT&T** American Telephone & Telegraph Company.

**AT&T 66** American Telephone & Telegraph 66 Cross Connect System.

**AT&T 110** American Telephone & Telegraph 110 Cross Connect System.

**ATM** Asynchronous transfer mode.

**Attenuation** The reduction in optical power as it passes along a fiber, usually expressed in decibels (dB); the reduction of signal strength over distance. See also optical loss.

**Attenuation coefficient** Characteristic of the attenuation of an optical fiber per unit length, in dB/km.

**Attenuator** A device that reduces signal power in a fiber optic link by inducing loss.

**AUI** Attachment unit interface.

**Average power** The average of a modulated signal over time.

**Backbone** Cable that connects communications closets, entrance facilities, and buildings.

**Back reflection, optical return loss** Light reflected from the cleaved or polished end of a fiber caused by the difference of refractive indices of air and glass: expressed in dB relative to incident power.

**Backscattering** The scattering of light in a fiber back toward the source; used to make OTDR measurements.

**Balanced pair transmission** Sending signals of opposite polarity on each wire in a pair to maximize bandwidth and minimize interference; used with all UTP cable.

**Balun** Balancing transformer; a device that adapts one cabling type to another, including physical layout, impedance, and connecting balanced to unbalanced cables.

**Bandwidth** The frequency spectrum required or provided by communications networks; the range of signal frequencies or bit rate within which a fiber optic component, link, or network will operate.

**Baud** For phone modems, it refers to the data rate, but in networks, it is the actual modulation rate that may not be the same as the data rate if encoding schemes are used.

**Baud rate** Rate of signal transmission; expressed in bps (bits per second).

**BBS** Backbone system/riser system.

**BD** Building distributor.

**BDN** Building distribution network.

**BDSL** Broadband digital subscriber line.

**Bending loss, microbending loss** Loss in fiber caused by stress on the fiber bent around a restrictive radius.

**Bend radius** Minimum radius a cable can be bent without permanent damage.

**BER** Bit error rate.

**Biconic** Conical molded 70 percent silica epoxy ferrule F/O connector designed by AT&T. One of the originals used by the TELCOs.

**Bit** Binary digit; an electrical or optical pulse that carries information; a single piece of digital information, a "1" or "0."

**Bit error rate (BER)** The fraction of data bits transmitted that are received in error.

**BIX** Northern Telecom's in-building cross-connect system.

**Block** A device used for interconnection of cables.

**BNC** Bayonet CXC connectors.

**Bonding** A permanent electrical connection.

**Bridge** A device that connects two or more sets of network wires.

**Buffer** A protective coating applied directly on the fiber.

**Bus** A network where all computers are connected by a single (usually coax) cable; bus architecture can also be implemented with a hub and star configuration.

**Byte** 8-bit binary word.

**Cable** One or more fibers enclosed in protective coverings and strength members.

**Cable plant, fiber optic** The combination of fiber optic cable sections, connectors, and splices forming the optical path between two terminal devices.

**Cable tray** A channel system used to hold and support communications cables.

**Capacitance** The ability of a conductor to store charge.

**Category 3** The UTP cable specified for signals up to 16 MHz, but commonly used for telephones.

**Category 4** The UTP cable specified for signals up to 20 MHz, but not commonly used for structured wiring systems.

**Category 5** The UTP cable specified for signals up to 100 MHz, commonly used for all LANs.

**CATV** An abbreviation for community antenna television, usually delivered by coax cable or HFC (hybrid fiber coax) networks, or cable TV.

**CBC** Communications building cable.

**CCITT** Consultants Committee for International Telephone and Telegraph.

**CCTV** Closed-circuit television, commonly used for security.

**CD** Campus distributor.

**CDDI** Copper distribution data interface.

**CEMA** Canadian Electrical Manufacturers Association.

**CFC** Communication flat cable.

**Chromatic dispersion** The temporal spreading of a pulse in an optical waveguide caused by the wavelength dependence of the velocities of light.

**Cladding** The lower refractive index optical coating over the core of the fiber that "traps" light into the core.

**Client** The computer that operates in a network using programs and data stored in a server.

**CM** Communications cables.

**CMG** Communications cable general rated.

**CMP** Communications cable plenum rated.

**CMR** Communications cable riser rated.

**CMX** Communications cable residential rated.

**Coax** A type of cable that uses a central conductor, insulation, outer conductor/shield, and jacket; used for high-frequency communications like CCTV or CATV.

**Coax, CXC** Coaxial cable.

**Conduit** A special pipe used to carry cables; may be metal or plastic, solid or flexible.

**Connector** The attachment on the end of a cable that allows interconnection to other cables.

**Connector, opitcal fiber** A device that provides for a demountable connection between two fibers or a fiber and an active device and provides protection for the fiber.

**Core** The center of the optical fiber through which light is transmitted.

**COSP** Customer-owned outside plant.

**Coupler** An optical device that splits or combines light from more than one fiber.

**CP** Consolidation point.

**CPE** Customer premises equipment.

**CPI** Component premises interface.

**Crimper** A tool used to install IDC plugs on cable.

**Crossed pair** A pair of wires in a UTP cable that have two pairs cross-connected in error.

**CSA** Canadian Standards Association.

**CSMA/CD** Carrier sensing multiple access/collision detection; the protocol of Ethernet and other networks using bus or star architecture, that controls access to the LAN.

**Current loop** Transmission using variable current to carry information, like a simple analog telephone.

**Cutback method** A technique for measuring the loss of bare fiber by measuring the optical power transmitted through a long length, then cutting back to the source and measuring the initial coupled power.

**Cutoff wavelength** The wavelength beyond which singlemode fiber only supports one mode of propagation.

**D4** Miniature, threaded, keyed F/O connectors.

**dB** Logarithmic ratio of signal levels, watts, voltage, or current; $dB = 10 \log 10 = (P_{out} / P_{in})$

**dBµ** Optical power referenced to 1 microwatt.

**dBm** In electrical usage, decibels in reference to milliwatts (0 dBm = 1 mW base measurement); optical power referenced to 1 milliwatt.

**DC** Direct current.

**DD** Distribution device.

**DEC** Digital Equipment Corporation.

**Decibel (dB)** A unit of measurement of optical power that indicates relative power on a logarithmic scale, sometimes called dBr; dB = 10 log (power ratio).

**Delay skew** The maximum difference of propagation time in all pairs of a cable.

**Detector** A photodiode that converts optical signals to electrical signals.

**Dial tone** The tone heard in a phone when the receiver is picked up, indicating the line is available for dialing.

**Dielectric** An insulator used to protect copper wires in cable.

**Digital** Signals in which information is in the form of digital bits, 1s and 0s, as opposed to analog signals.

**Dispersion** The temporal spreading of a pulse in an optical waveguide; may be caused by modal or chromatic effects.

**DIW** "D" inside wire.

**DMARC** Demarcation point.

**DNP** Dry Nonpolish F/O connectors for POF.

**DOC** Department of Communications.

**DS0** Digital system level "0," 64 KB/s.

**DS1** Digital system level "1," 1.544 MB/s.

**DS2** Digital system level "2," 6.312 MB/s.

**DS3** Digital system level "3," 44.736 MB/s.

**DTE** Data terminal equipment.

**DTMF** Dual-tone multifrequency, or tone dialing used on modern phones, where discrete tones indicate numbers.

**DVO** Data voice outlet.

**EDFA** Erbium-doped fiber amplifier, an all-optical amplifier for 1,550 nm SM transmission systems.

**Edge-emitting diode (E-LED)** An LED that emits from the edge of the semiconductor chip, producing higher power and narrower spectral width.

**EIA/TIA** Electronics Industry Association/Telecommunications Industry Association; a vendor-based group that writes voluntary interoperability standards for communications and electronics.

**EIA/TIA 568 standard** A voluntary standard developed by vendors to ensure interoperability of equipment used on network cabling.

**EL-FEXT** Equal level far end crosstalk; crosstalk at the far end with signals of equal level being transmitted.

**EMD** Equilibrium mode distribution.

**EMI** Electromagnetic interference.

**End finish** The quality of the end surface of a fiber prepared for splicing or terminated in a connector.

**Equilibrium modal distribution** (EMD) Steady state modal distribution in multimode fiber, achieved some distance from the source, where the relative power in the modes becomes stable with increasing distance.

**ESCON** Enterprise System Connection; IBM standard for connecting peripherals to a computer over fiber optics.

**ESD** Electrostatic discharge.

**Ethernet** A 10 MB/s LAN that is based on bus or hub architecture.; most widely used LAN.

**ETL** Electrical Test Laboratories.

**Excess loss** The amount of light lost in a coupler, beyond that inherent in the splitting to multiple output fibers.

**FC** Threaded, keyed F/O connectors.

**FCC** Federal Communications Commission; oversees all communications issues in the United States.

**FD** Floor distributor.

**FDDI** Fiber distributed data interface; 100 MB/s Token Ring network developed for optical fiber.

**FDM** Frequency division multiplexing.

**Ferrule** A precision tube that holds a fiber for alignment for interconnection or termination; may be part of a connector or mechanical splice.

**Fiber amplifier** An all-optical amplifier using erbium or other doped fibers and pump lasers to increase signal output power without electronic conversion.

**Fiber distributed data interface** (FDDI) 100 MB/s ring architecture data network.

**Fiber identifier** A device that clamps onto a fiber and couples light from the fiber by bending, to identify the fiber and detect high-speed traffic of an operating link or a 2-kHz tone injected by a test source.

**Fiber optics** Light transmission through flexible transmissive fibers for communications or lighting.

**Fiber tracer** An instrument that couples visible light into the fiber to allow visual checking of continuity and tracing for correct connections.

**Firestop** Restores a fire-rated partition to its fire rating after penetration with cabling.

**Fishtape** Semiflexible rod used to retrieve cables or pull line.

**FLAT** Flat cable and under-carpet cable.

**FMPR** Fiber optic multiport repeater.

**FO** Common abbreviation for "fiber optic" or "fiber optics."

**F/O** Fiber optic cable.

**FOFOU** Fiber optic fan out line.

**FOH** Fiber optic hub.

**FOR** Fiber optic repeater.

**FOT** Fiber optic transceiver.

**FOTP** Fiber optic test procedures.

**FOTS** Fiber optic transmission system.

**Fresnel reflection, back reflection, optical return loss** Light reflected from the cleaved or polished end of a fiber caused by the difference of refractive indices of air and glass; typically 4 percent of the incident light.

**FSD** Fixed Shroud Duplex F/O Connector; AMP trade name for FDDI connector.

**F Series Connector** Threaded RG59 CXC connector.

**FSMA/SMA** Fixed shroud multiple adaptation F/O connector.

**FT1(Can)** Vertical flame test for general-purpose cables in combustible buildings (same as CMX or CMG rating).

**FT4(Can)** Vertical flame test for cables in tray in noncombustible buildings in riser applications (same as CMR rating).

**FT6(Can)** Horizontal flame and smoke test for cables in noncombustible building in plenum applications (same as CMP Rating).

**FTTD** Fiber to the desk.

**FTTH** Fiber to the home.

**Fusion splicer** An instrument that splices fibers by fusing or welding them, typically by electrical arc.

**GBE** Gigabit Ethernet.

**GHz** Gigahertz.

**Graded index (GI)** A type of multimode fiber that uses a graded profile of refractive index in the core material to correct for dispersion.

**Ground** A connection between a circuit or equipment and the earth.

**Ground loop** The flow of current caused by unequal ground potentials.

**HC** Horizontal cross-connect.

**HCS** Hard clad silica fiber.

**HDS** Horizontal distribution system.

**HDSL** High bit rate digital subscriber line.

**Headend** The main distribution point in a CATV system.

**HFC** Hybrid fiber coax CATV network.

**HIPPI** High-performance parallel interface.

**Horizontal** Cable that runs from a device to the communications closet.

**Horizontal cross-connect** Connection of horizontal wiring to other equipment or cabling.

**Host** Large computer used with terminals, usually a mainframe.

**Hub** A switch used to connect computers in a star network.

**HYBRID** Mixed media conductor cables; in fiber optics, refers to cable with both singlemode and multimode fibers.

**Hz** Frequency; cycles per second.

**IBDN** Integrated Building Distribution Network; NT cabling system most recently owned by CDT.

**IBM** International Business Machines.

**IBM 3270** Large mainframe computer family.

**IC** Intermediate cross-connect.

**ICEA** Insulated Cable Engineers Association.

**ICS** The IBM cabling system.

**IDC** Insulation displacement connection (connector), connecting wires by inserting or crimping cable into metal contacts that cut through the insulation, making contact with the wires.

**IDF** Intermediate distribution frame.

**IEC** International Electrotechnical Committee; oversees international communications standards.

**IEEE** Institute of Electrical and Electronics Engineers; professional society that oversees network standards.

**Impedance** The AC resistance.

**Impedance matching devices** A type of balun that matches impedance between two cables.

**Index matching fluid** A liquid of refractive index similar to glass used to match the materials of two fibers to reduce loss and back reflection.

**Index of refraction** A measure of the speed of light in a material.

**Index profile** The refractive index of a fiber as a function of cross section.

**Insertion loss** The loss caused by the insertion of a component such as a splice or connector in an optical fiber.

**Intermediate cross-connect** Connection point in the backbone cable between the main cross-connect and the telecommunications closet.

**Internet** A worldwide network of computers that allows communications from computers.

**IRL** Interrepeater link.

**ISDN** Integrated services digital network.

**ISO** International Standards Organization; oversees international standards.

**Jacket** The outer protective covering of a cable.

**J hook** A hook shaped like the letter "J" used to suspend cables.

**Jumper cable** A short single fiber cable with connectors on both ends used for interconnecting other cables or testing.

**Key system** A simple multiline phone system that allows each user to select from several lines.

**km** Kilometer.

**LAN** Local area network; a group of computers and peripherals set up to communicate with each other.

**LASER** Light amplification by stimulated emission of radiation.

**Laser diode (ILD)** A semiconductor device that emits high-powered, coherent light when stimulated by an electrical current; used in transmitters for singlemode fiber links.

**Launch cable** A known good fiber optic jumper cable attached to a source and calibrated for output power used for loss testing. This cable must be made of fiber and connectors of a matching type to the cables to be tested.

**Light Emitting Diode (LED)** A semiconductor device that emits light when stimulated by an electrical current; used in transmitters for multimode fiber links.

**Link, fiber optic** A combination of transmitter, receiver, and fiber optic cable connecting them capable of transmitting data; may be analog or digital.

**Local loop** The interconnection of telephone central offices in a small region.

**Long wavelength** A commonly used term for light in the 1,300- and 1,550-nm ranges.

**Loop resistance** A measurement of the resistance of both wires in a pair measured from one end with the other end shorted.

**Loss budget** The amount of power lost in the link; often used in terms of the maximum amount of loss that can be tolerated by a given link.

**Loss, optical** The amount of optical power lost as light is transmitted through fiber, splices, couplers, and so on.

**MAC** Media access control.

**Main cross-connect** The connection point between building entrance, backbone and equipment cables.

**Mainfame** A large computer used to store and process massive amounts of data.

**MAN** Metropolitan area network.

**Margin** The additional amount of loss that can be tolerated in a link.

**MAU** Media attachment unit.

**MB/s** Megabits per second.

**MC** Main cross-connect.

**MDF** Main distribution frame.

**Mechanical splice** A semipermanent connection between two fibers made with an alignment device and index matching fluid or adhesive.

**Mesh grip (Kellums grip)** A grip made of wire mesh that grips the jacket of a cable for pulling.

**Messenger cable** The aerial cable used to attach communications cable that has no strength member of its own.

**MHz** Megahertz; millions of cycles per second.

**MIC** Media interface connector (usually refers to the FDDI connector).

**Micron (µm)** A unit of measure, $10^{-6}$ m, used to measure wavelength of light.

**Microscope, fiber optic inspection** A microscope used to inspect the end surface of a connector for flaws or contamination or a fiber for cleave quality.

**MM** Multimode (fiber).

**MMJ** Modified modular jack.

**MMP** Modified modular plug.

**Modal dispersion** The temporal spreading of a pulse in an optical waveguide caused by modal effects.

**Mode** A single electromagnetic field pattern that travels in fiber.

**Mode field diameter** A measure of the core size in singlemode fiber.

**Mode filter** A device that removes optical power in higher-order modes in fiber.

**Modem** Modulator/demodulator.

**Mode scrambler** A device that mixes optical power in fiber to achieve equal power distribution in all modes.

**Mode stripper** A device that removes light in the cladding of an optical fiber.

**Modular 8** The proper name for the 8-pin connector used in EIA/TIA 568 standard; commonly called RJ-45.

**Modular jack** A female connector for wall or panel installation; mates with modular plugs.

**Modular plug** A standard connector used with wire, with 4 to 10 contacts, to mate cables with modular jacks.

**MP** Multipurpose (cables).

**MSAU** Multistation access unit.

**Multimode fiber** A fiber with core diameter much larger than the wavelength of light transmitted that allows many modes of light to propagate; commonly used with LED sources for lower-speed, short-distance links.

**NA** Numerical aperture; a measure of the angular acceptance of an optical fiber.

**Nanometer (nm)** A unit of measure, $10^{-9}$ m, used to measure the wavelength of light.

**NCTA** National Cable Television Association.

*NEC®* *National Electrical Code®*, written by NFPA, this code sets standards for fire protection for construction.

**NECA** National Electrical Contractors Association.

**NEMA** National Electrical Manufacturers Associations.

**Network** A system of cables, hardware, and equipment used for communications.

**Network interface (NI)** The demarcation point where the public network connects to a private (commercial or residential) network.

**NEXT** Near end crosstalk; a measure of interference between pairs in UTP cable.

**NFPA** National Fire Protection Association, which writes the *NEC®*.

**NIC** Network interface card; used to interface computers to networks.

**NII** National information infrastructure.

**NIST** National Institute of Standards and Technology; establishes primary standards in the United States.

**nm** Nanometer; one - one billionth of a meter.

**NMS** Network management system.

**NOS** Network operating system; the software that allows computers on a network to share data and program files.

**N Series Connector** Threaded Thicknet CXC connector.

**NT** Northern Telecom.

**NTT** Nippon Telephone & Telecommunications.

**Numerical aperture (NA)** A measure of the light acceptance angle of the fiber.

**NVP** Nominal velocity of propagation; that speed of signal travel in the cable, expressed relative to the speed of light.

**OBC** Ontario Building Code (Canada).

**OC1** Optical Capacity - 1, 51.8 MB/s (OC12 = 12 x 51.8 = 622 MB/s ).

**OFC** Optical fiber cable conductive general rated.

**OFCP** Optical fiber cable conductive plenum rated.

**OFCR** Optical fiber cable conductive riser rated.

**OFN** Optical fiber cable nonconductive general rated.

**OFNP** Optical fiber cable nonconductive plenum rated.

**OFNR** Optical fiber cable nonconductive riser rated.

**OHESC** Ontario Hydro Electrical Safety Code (Canada).

**Ohm** Standard unit of electrical resistance.

**OPM** Optical power meter.

**Optical amplifier** A device that amplifies light without converting it to an electrical signal.

**Optical fiber** An optical waveguide, comprised of a glass or plastic light carrying core and cladding, that traps light in the core and can carry communications signals.

**Optical loss** a reduction of optical power caused by the attenuation of fiber, connectors, and splices.

**Optical loss test set (OLTS)** A measurement instrument for optical loss that includes both a meter and source.

**Optical power** The amount of radiant energy per unit time, expressed in linear units of watts or on a logarithmic scale, in dBm (where 0 dB = 1 mW) or dBμ (where 0 dBμ = 1 mW).

**Optical return loss, back reflection** Light reflected from the cleaved or polished end of a fiber caused by the difference of refractive indices of air and glass; typically 4 percent of the incident light; expressed in dB relative to incident power.

**Optical switch** A device that routes an optical signal from one or more input ports to one or more output ports.

**Optical time domain reflectometer (OTDR)** An instrument that uses backscattered light to find faults in optical fiber and infer loss.

**OSHA** Occupational Safety and Health Administration.

**OSI** Open systems interconnect.

**OTDR** Optical time domain reflectometer.

**Overfilled launch** A condition for launching light into the fiber where the incoming light has a spot size and NA larger than accepted by the fiber, filling all modes in the fiber.

**PANS** Pretty amazing new stuff.

**Patch panel** A cross-connection using jacks and patchcords to interconnect cables.

**PBX or PABX** Private branch exchange or private automatic branch exchange.

**PCC** Premises communications cable.

**PCI** Peripheral component interconnect.

**PCM** Pulse-coded modulation.

**PCS** Plastic-clad silica.

**PDS** Premises distribution system.

**PE** Polyethylene.

**Photodiode** A semiconductor that converts light to an electrical signal; used in fiber optic receivers.

**PIC** Plastic insulated conductor.

**Pigtail** A short length of fiber attached to a fiber optic component such as a laser or coupler.

**PIN Photodiode** a fast photo detector used in many fiber optic receivers

**Plastic-clad silica (PCS) fiber** A fiber made with a glass core and plastic cladding.

**Plastic optical fiber (POF)** An optical fiber made of plastic.

**Plenum** The air-carrying portion of a heating or air-conditioning system that can be used for running communications cables; also a type of cable used in plenums, specially rated by the *NEC*®.

**PNI** Premises network interface.

**POTS** Plain old telephone service or plain ordinary telephone systems.

**Power budget** The difference (in dB) between the transmitted optical power (in dBm) and the receiver sensitivity (in dBm).

**Power meter, fiber optic** An instrument that measures optical power emanating from the end of a fiber.

**Power Sum NEXT** Near end crosstalk tested with all pairs but one energized to find the total amount of crosstalk caused by simultaneous use of all pairs for communication.

**Preform** The large-diameter glass rod from which fiber is drawn.

**PTSS** Passive transmission subsystem.

**Pulse dialing** Old-style phone dialing that works by making and breaking the current loop a number of times to indicate the number dialed.

**Punchdown block** A connection block incorporating insulation displacement connections for interconnecting copper wires with a special insertion tool.

**Punchdown tool** A tool used to connect wire to IDC connections in punchdown blocks.

**PUR** Polyurethane.

**PVC** Polyvinyl chloride.

**QCBIX1A** 25-pair punchdown block, 5-pair grouping.

**QCBIX1A4** 25-pair punchdown block, 4-pair grouping.

**QMBIX10A** BIX mounting frame.

**QRBIX19A** Distribution ring.

**QSBIX20A** Field designation strip.

**QTBIX16A** Punchdown tool.

**QUAD** Four-conductor cable.

**REA** Rural Electrification Association.

**Receive cable** A known good fiber optic jumper cable attached to a power meter used for loss testing. This cable must be made of fiber and connectors of a matching type to the cables to be tested.

**Receiver** A device containing a photodiode and signal conditioning circuitry that converts light to an electrical signal in fiber optic links.

**Refractive index** A property of optical materials that relates to the velocity of light in the material.

**Repeater, regenerator** A device that receives a fiber optic signal and regenerates it for retransmission; used in very long fiber optic links.

**Return loss** Reflection from an impedance mismatch in a copper cable.

**Reversed pair** A pair of wires in a UTP cable that has the two wires cross-connected in error.

**RFI** Radio frequency interference.

**RFP** Request for proposal.

**RFQ** Request for quotation.

**RG** Residential gateway.

**Ring** One conductor in a phone line, connected to the "Ring" of the contact on old-fashioned phone plugs; a network where computers are connected in series to form a ring—each computer in turn has an opportunity to use the network.

**RJ-11** 6-position modular jack/plug.

**RJ-45** A modular 8-pin connector, actually referring to a specific telephone application, but usually referring to the connector used in the EIA/TIA 568 standard.

**RSD** Retractable shroud duplex F/O connector.

**SC** Connector square format optical fiber connector, push-pull type; EIA/TIA standard, NTT design, miniature Push-pull F/O connectors.

**Scattering** The change of direction of light after striking small particles that causes loss in optical fibers.

**SCTE** Society of Cable and Telecommunications Engineers.

**ScTP** screened twisted pair cable; UTP cable with a outer shield under the jacket to prevent interference.

**Server** The center of a network where programs and data are stored.

**SFF** Small form factor, for example, new small form factor fiber optic connectors.

**Shorted pair** A pair of wires in a UTP cable that are electrically connected in error.

**Short wavelength** A commonly used term for light in the 665-, 790-, and 850-nm ranges.

**Singlemode fiber** A fiber with a small core, only a few times the wavelength of light transmitted, that allows only one mode of light to propagate; commonly used with laser sources for high-speed, long-distance links.

**SM** Singlemode (fiber).

**SNR** Signal-to-noise ratio.

**SONET** Synchronized optical network.

**Source** A laser diode or LED used to inject an optical signal into fiber.

**Splice (fusion or mechanical)** A device that provides for a connection between two fibers, typically intended to be permanent.

**Split pair** A pair of wires in a UTP cable that have the two wires of two different pairs cross-connected in error.

**Splitting ratio** The distribution of power among the output fibers of a coupler.

**SRL** Structural return loss.

**ST** Keyed, 2.5-mm, ferrule bayonet F/O connectors.

**Star** A network in which all the computers are connected to a central hub or server.

**Steady state modal distribution** Equilibrium modal distribution (EMD) in multimode fiber, achieved some distance from the source, where the relative power in the modes becomes stable with increasing distance.

**Step index fiber** A multimode fiber where the core is all the same index of refraction.

**STP** Shielded twisted pair cable, where each pair has a metallic shield to prevent interference.

**Structured cabling** A method of installing cable per industry standards to allow interoperability among vendors and upgrades.

**Subscriber loop** Connection of the end user to the local central office telephone switch.

**Surface emitter LED** An LED that emits light perpendicular to the semiconductor chip. Most LEDs used in data communications are surface emitters.

**Systimax PDS** AT&T wiring system.

**T1** Transmission Level 1, first level of multiplexing, 1.554 MB/s, (Bell standard).

**T568A** 4-pair EIA/TIA modular plug wiring scheme, ISDN.

**T568B** 4-pair EIA/TIA modular plug wiring scheme, AT&T.

**Takeoff** Reading drawings of a layout to get a cable plant layout.

**Talkset, fiber optic** A communication device that allows conversation over unused fibers.

**TC** Telecommunications closet.

**TDM** Time division multiplexing.

**TDR** Time domain reflectometer.

**TELCO** Telephone Company (Bell, MCI, and so on).

**Telecommunications closet** Location inside a building for interconnection of backbone and horizontal cables.

**Telegraph** The earliest form of long-distance communications, using coded letters.

**Telephone** A voice instrument for communications.

**Telephone switch** A device that connects telephones together when signaled by dialing.

**Termination** Preparation of the end of a fiber to allow connection to another fiber or an active device; sometimes called connectorization.

**Test cable** A short single fiber jumper cable with connectors on both ends used for testing. This cable must be made of fiber and connectors of a matching type to the cables to be tested.

**Test kit** A kit of fiber optic instruments, typically including a power meter, source, and test accessories, used for measuring loss and power.

**Test source** A laser diode or LED used to inject an optical signal into fiber for testing loss of the fiber or other components.

**TIA** Telecommunications Industries Association.

**TIC** Token Interface Card, 802.5 Token Ring.

**Time domain reflectometer (TDR)** A testing device used for copper cable that operates like radar to find length, shorts or opens, and impedance mismatches.

**Tip** One conductor in a phone line, connected to the "Tip" of the old-fashioned phone plug.

**TLA** Three-letter acronym.

**TO** Telecommunications outlet/connector.

**TOC** Telecommunications outlet connector.

**Token Ring** A ring architecture LAN developed by IBM; 4 MB/s and 16 MB/s versions are used.

**Tone dialing** Used on modern phones, where discrete tones indicate numbers.

**Topology** The architecture or layout of a network (for example, bus, ring, star).

**Total internal reflection** Confinement of light into the core of a fiber by the reflection off the core-cladding boundary.

**TP** Transition point.

**TPDDI** Twisted pair distribution data interface.

**TPPMD** Twisted pair physical media dependent.

**Transmitter** A device that includes an LED or laser source and signal conditioning electronics that is used to inject a signal into fiber.

**TSB** Technical Service Bulletin.

**UL** Underwriters Laboratories (United States).

**ULC** Underwriters Laboratories Canada.

**UN** Universal network interface.

**USOC** Uniform Service Order Code; a UTP wiring scheme that allows 6-pin plugs to be used in 8-pin jacks for telephone use.

**UTP** unshielded twisted pair cable; comprised of four pairs of conductors carefully manufactured to preserve frequency characteristics.

**VAR** Value added reseller (Retailer).

**VCSEL** Vertical cavity surface emitting laser; a type of laser that emits light vertically out of the chip, not out the edge.

**VDSL** Very high bit rate digital subscriber line.

**VESA** Video Electronic Standards Association.

**Visual fault locator** A device that couples visible light into the fiber to allow visual tracing and testing of continuity. Some are bright enough to allow finding breaks in fiber through the cable jacket.

**WAN** Wide area network.

**Watts** A linear measure of optical power, usually expressed in milliwatts (mW), microwatts (μW), or nanowatts (nW).

**Wavelength** A measure of the color of light, usually expressed in nanometers (nm) or microns (μm).

**Wavelength division multiplexing (WDM)** A technique of sending signals of several different wavelengths of light into the fiber simultaneously.

**Wireless** Sending communications over radio waves.

**Wire mapping** Confirming the proper connections of all four pairs of UTP cabling.

**Work area** The location of the equipment connected to horizontal cabling; sometimes called the desktop.

**Work area outlet** The outlet at the end of the horizontal cabling where equipment is connected with a patchcord.

**Working margin** The difference (in dB) between the power budget and the loss budget (i.e., the excess power margin).

**World Wide Web** The graphical communication network operating over the Internet.

**YAA** Yet another acronym?

**Z, Zo** Impedance symbol.

**ZSW** "Z" station wire or quad cable (same as QUAD).

# Appendix A: Labor Rates for Communications Wiring

*(Labor Units in Hours)*

## CCTV/CATV

| Television Systems | Normal | Difficult | Very Difficult |
|---|---|---|---|
| video camera | 0.80 | 1.00 | 1.20 |
| power bracket | 0.50 | 0.65 | 0.80 |
| camera control unit | 0.80 | 1.00 | 1.20 |
| standard mounting | 0.60 | 0.75 | 0.90 |
| swivel camera mount | 0.70 | 0.90 | 1.10 |
| tamper-proof mounting | 1.50 | 2.00 | 2.40 |
| automatic pan mounting | 1.50 | 2.00 | 2.40 |
| add or replace lens | 0.40 | 0.50 | 0.60 |

| Monitors | Normal | Difficult | Very Difficult |
|---|---|---|---|
| 9" monitor | 0.90 | 1.15 | 1.40 |
| 12" monitor | 1.00 | 1.25 | 1.50 |
| 15" monitor | 1.10 | 1.40 | 1.70 |
| 19" monitor | 1.20 | 1.50 | 1.80 |
| monitor mounting bracket | 1.00 | 1.25 | 1.50 |

| Switches and Sequencers | Normal | Difficult | Very Difficult |
|---|---|---|---|
| 4 camera switch, manual | 1.20 | 1.50 | 1.80 |
| 8 camera switch, manual | 1.50 | 2.00 | 2.40 |
| 4 camera sequencer | 1.50 | 2.00 | 2.40 |
| 8 camera sequencer | 1.80 | 2.25 | 2.70 |

| Miscellaneous | Normal | Difficult | Very Difficult |
|---|---|---|---|
| TV receptacle | 0.30 | 0.40 | 0.50 |
| 2-wire plug or connector | 0.20 | 0.25 | 0.30 |
| 4-wire plug or connector | 0.30 | 0.40 | 0.50 |
| amplifier | 1.00 | 0.20 | 1.50 |
| coupler | 0.60 | 0.75 | 0.90 |
| splitter | 0.60 | 0.75 | 0.90 |
| coax surge suppressor | 0.75 | 1.00 | 1.75 |
| video motion detector | 1.00 | 1.25 | 1.50 |
| screen splitter | 0.60 | 0.75 | 0.90 |
| head end equipment | 6.00 | 10.00 | 14.00 |
| antenna | 8.00 | 10.00 | 12.00 |
| console | 6.00 | 9.00 | 12.00 |
| satellite receiver, 4’ | 10.00 | 13.00 | 16.00 |
| satellite receiver, 6’ | 12.00 | 15.00 | 18.00 |
| satellite receiver, 8’ | 16.00 | 20.00 | 24.00 |
| 19” rack adaptors | 0.50 | 0.65 | 0.80 |
| 19” rack, 2’ | 1.80 | 2.25 | 2.70 |
| 19” rack, 4’ | 3.00 | 3.75 | 4.50 |
| 19” rack, 6’ | 4.50 | 6.00 | 7.50 |
| tie-wraps | 0.01 | 0.02 | 0.03 |
| wire markers | 0.01 | 0.02 | 0.03 |

| Cables (per ft) Accessible Locations | Normal | Difficult | Very Difficult |
|---|---|---|---|
| coaxial | 0.012 | 0.015 | 0.018 |
| coaxial crimp connectors | 0.08 | 0.10 | 0.12 |
| coaxial twist-on connectors | 0.08 | 0.10 | 0.12 |
| coaxial tee connectors | 0.10 | 0.13 | 0.16 |
| twinaxial connectors | 0.15 | 0.2 | 0.25 |

## COMPUTERS AND NETWORKS

| Computer Labor Units | Normal | Difficult | Very Difficult |
|---|---|---|---|
| Computer (PC), processor only | 1.00 | 1.50 | 2.00 |
| Computer (PC), with monitor | 1.50 | 2.00 | 2.50 |
| Computer, minicomputer (AS400) | 6.00 | 9.00 | 12.00 |
| Computer, mainframe | 12.00 | 18.00 | 24.00 |

| Peripheral Items | Normal | Difficult | Very Difficult |
|---|---|---|---|
| external modem | 0.60 | 0.75 | 0.90 |
| modem sharing kit | 1.00 | 1.50 | 2.00 |
| desktop scanner | 1.00 | 1.25 | 1.50 |
| external tape backup | 0.80 | 1.00 | 1.20 |
| bar code reader, handheld | 1.00 | 1.25 | 1.50 |

| Miscellaneous | Normal | Difficult | Very Difficult |
|---|---|---|---|
| modular cables | 0.20 | 0.25 | 0.30 |
| cable adapters | 0.30 | 0.40 | 0.50 |

| Network Devices | Normal | Difficult | Very Difficult |
|---|---|---|---|
| network interface card | 1.00 | 1.25 | 1.50 |
| host adapter, with cable | 1.00 | 1.25 | 1.50 |
| LAN hub concentrater | 2.00 | 4.00 | 6.00 |
| multiplexer | 2.50 | 5.00 | 8.00 |
| telecluster | 4.00 | 6.00 | 8.00 |
| Ethernet adapter | 0.50 | 0.65 | 0.80 |
| wireless transceiver | 3.00 | 3.75 | 4.50 |
| long-distance adapter | 1.00 | 2.00 | 3.00 |
| signal booster | 1.00 | 1.50 | 2.00 |
| line splitter | 2.00 | 2.50 | 3.00 |
| bridge | 4.00 | 6.00 | 8.00 |
| router | 4.00 | 6.00 | 8.00 |
| WAN interface card | 4.00 | 6.00 | 8.00 |
| repeater | 4.00 | 6.00 | 8.00 |
| diagnostic modem | 2.00 | 4.00 | 6.00 |

| Printers | Normal | Difficult | Very Difficult |
|---|---|---|---|
| dot matrix printer | 1.00 | 1.50 | 2.00 |
| laser printer | 1.20 | 1.70 | 2.20 |
| color printer | 1.20 | 2.00 | 2.50 |
| emulation card | 1.00 | 1.25 | 1.50 |
| A/B printer switch, one cable | 1.00 | 1.25 | 1.50 |
| multiprinter box | 1.50 | 2.00 | 2.40 |
| printer buffer, external | 1.00 | 1.25 | 1.50 |
| pen plotter | 3.00 | 4.00 | 5.00 |

**Network cables (per ft)**

| Accessible Locations | Normal | Difficult | Very Difficult |
|---|---|---|---|
| unshielded twisted pair | 0.01 | 0.013 | 0.015 |
| shielded twisted pair | 0.01 | 0.013 | 0.015 |
| coaxial | 0.012 | 0.015 | 0.018 |
| 4-conductor telephone cable | 0.01 | 0.013 | 0.015 |
| other types | 0.012 | 0.015 | 0.018 |

| Miscellaneous, Cable Related | Normal | Difficult | Very Difficult |
|---|---|---|---|
| coaxial crimp connectors | 0.08 | 0.10 | 0.12 |
| coaxial twist-on connectors | 0.08 | 0.10 | 0.12 |
| coaxial tee connectors | 0.10 | 0.12 | 0.15 |
| twinaxial connectors | 0.15 | 0.2 | 0.25 |
| shielded twisted pair terminations | 0.12 | 0.15 | 0.18 |
| unshielded twisted pair terminations | 0.1 | 0.13 | 0.16 |
| punchdown block, no terminations | 0.8 | 1.00 | 1.20 |
| wallplate with jack | 0.50 | 0.65 | 0.8 |
| modular connectors, telephone | 0.15 | 0.2 | 0.25 |
| in-line couplers | 0.10 | 0.13 | 0.16 |
| male/female adapters | 0.10 | 0.13 | 0.16 |
| cable end asemblies | 0.60 | 0.75 | 0.90 |
| extension | 0.20 | 0.25 | 0.30 |

## TELEPHONES

| Telephones | Normal | Difficult | Very Difficult |
|---|---|---|---|
| Single-line desk phone | 0.20 | 0.25 | 0.30 |
| 2-line desk phone | 0.24 | 0.30 | 0.36 |
| 4- to 8-line desk phone | 0.30 | 0.40 | 0.50 |
| central console | 1.50 | 2.00 | 2.40 |

| Large Systems | Normal | Difficult | Very Difficult |
|---|---|---|---|
| auto switching unit | 4.00 | 5.00 | 6.00 |
| remote control for above | 2.00 | 2.50 | 3.00 |
| 8-line central unit | 6.00 | 9.00 | 12.00 |
| 16-line central unit | 8.00 | 12.00 | 16.00 |
| 24-line central unit | 10.00 | 14.00 | 18.00 |
| 40-line central unit | 16.00 | 24.00 | 32.00 |
| 80-line central unit | 25.00 | 32.00 | 40.00 |
| digital line card | 0.50 | 0.65 | 0.75 |

| Cables (per ft) Accessible Locations | Normal | Difficult | Very Difficult |
|---|---|---|---|
| unshielded twisted pair | 0.01 | 0.013 | 0.015 |
| shielded twisted pair | 0.01 | 0.013 | 0.015 |
| coaxial | 0.012 | 0.015 | 0.018 |
| 4-conductor telephone cable | 0.01 | 0.013 | 0.015 |
| other types | 0.012 | 0.02 | 0.02 |

| Miscellaneous, Cable Related | Normal | Difficult | Very Difficult |
|---|---|---|---|
| shielded twisted pair terminations | 0.12 | 0.15 | 0.18 |
| unshielded twisted pair terminations | 0.10 | 0.13 | 0.15 |

| | | | |
|---|---|---|---|
| punchdown block, no terminations | 1.00 | 1.25 | 1.50 |
| wallplate with jack | 0.80 | 1.00 | 1.20 |
| modular connectors, telephone | 0.15 | 0.20 | 0.25 |
| in-line couplers | 0.10 | 0.13 | 0.15 |
| male/female adapters | 0.40 | 0.50 | 0.60 |
| cable end asemblies | 0.60 | 0.75 | 0.90 |
| extension | 0.20 | 0.25 | 0.30 |
| punchdown block, no terminations | 1.00 | 1.25 | 1.50 |
| modular telephone connectors | 0.15 | 0.20 | 0.25 |
| 19” rack, 2’ | 1.80 | 2.25 | 2.70 |
| 19” rack, 4’ | 3.00 | 3.75 | 4.50 |
| 19” rack, 6’ | 4.50 | 5.70 | 6.75 |
| tie-wraps | 0.01 | 0.02 | 0.03 |
| wire markers | 0.01 | 0.02 | 0.03 |

## FIBER OPTICS

| Optical Fiber Cables (per ft) | Normal | Difficult |
|---|---|---|
| 1 to 4 fibers, in conduit | 0.016 | 0.02 |
| 1 to 4 fibers, accessible locations | 0.014 | 0.018 |
| 12 to 24 fibers, in conduit | 0.02 | 0.025 |
| 12-24 fibers, accessible locations | 0.018 | 0.023 |
| 48 fibers, in conduit | 0.03 | 0.038 |
| 48 fibers, accessible locations | 0.025 | 0.031 |
| 72 fibers, in conduit | 0.04 | 0.05 |
| 72 fibers, accessible locations | 0.032 | 0.04 |
| 144 fibers, in conduit | 0.05 | 0.065 |
| 144 fibers, accessible locations | 0.04 | 0.05 |
| **Hybrid Cables** | | |
| 1 to 4 fibers, in conduit | 0.02 | 0.025 |
| 1 to 4 fibers, accessible locations | 0.017 | 0.021 |
| 12 to 24 fibers, in conduit | 0.024 | 0.03 |
| 12 to 24 fibers, accessible locations | 0.022 | 0.028 |
| **Testing (per fiber)** | 0.12 | 0.24 |
| **Splices (inc. prep time and failures) Trained Workers** | | |
| fusion | 0.30 | 0.45 |
| mechanical | 0.40 | 0.50 |
| array splice, 12 fibers | 1.00 | 1.30 |
| **Coupler (connector-connector)** | 0.15 | 0.25 |

| **Terminations (inc. prep time and failures)** | **Trained Workers** | |
|---|---|---|
| polishing required | 0.40 | 0.60 |
| no-polish connectors | 0.30 | 0.45 |
| FDDI dual connector, including terminations | 0.80 | 1.00 |
| **Miscellaneous** | | |
| cross-connect box, 144 fibers (not inc. splices) | 3.00 | 4.00 |
| splice cabinet | 2.00 | 2.50 |
| splice case | 1.80 | 2.25 |
| breakout kit, 6 fibers | 1.00 | 1.40 |
| tie-wraps | 0.01 | 0.02 |
| wire markers | 0.01 | 0.01 |

# Appendix B: Standards Groups

ANSI
American National Standards Institute
11 W. 42nd St.
New York, NY 10036
212-642-4900

Bellcore
Information Exchange Management
445 South St.
Box 1910
Morristown, NJ 07962-1910
201-829-4785

EIA/TIA
Electronic Industry Association
1722 Eye St. N.W., Suite 440
Washington, DC 20006
202-457-4912

*Most standards can be purchased from:*
Global Engineering Documents
15 Inverness Way E.
Englewood, CO 80112-5704
303-792-2181

IEEE Standards
445 Hoes Lane
Piscataway, NJ 08855
908-562-3800

International Electrotechnical Commission
PO Box 131
3 Rue de Varembe
1211 Geneva 20
Switzerland
41-22-734-0150

United States Department of Defense
DESC-EMT
Dayton, OH 45444-5284
513-296-5541

# INDEX